Industry 4.0

This book presents a comprehensive discussion of the recent advances in Industry 4.0, manufacturing processes, and intelligent techniques. It will serve as an ideal reference text for graduate students and academic researchers in the fields of manufacturing engineering, industrial engineering, mechanical engineering, and production engineering. This text introduces Industry 4.0, its evolution, and essential pillars of Industry 4.0 including calibration, metrology, quality control, robotics, artificial intelligence, and the Internet of Things.

It comprehensively covers important topics including the cold spray technique for additive manufacturing, tool condition monitoring, robotic manipulators, metrology, quality control, and the Internet of Things in Industry 4.0.

The book

- Discusses additive manufacturing and applications of lasers in advanced manufacturing.
- Covers sensors, actuators, and calibration techniques for next-generation industries.
- Emphasizes the recycling of materials for sustainable manufacturing.
- Explores latest advances in the Internet of Things, robotics, artificial intelligence, and machine learning in view of Industry 4.0.
- Provides a conceptual framework of Industry 4.0 with the help of applications and case studies.

The text is primarily written for graduate students and academic researchers in the fields of manufacturing engineering, industrial engineering, mechanical engineering, and production engineering.

Industry 4.0

Concepts, Processes and Systems

Edited by

Ravi Kant
Hema Gurung

CRC Press
Taylor & Francis Group
Boca Raton London New York

CRC Press is an imprint of the
Taylor & Francis Group, an **informa** business

First edition published 2024
by CRC Press
6000 Broken Sound Parkway NW, Suite 300, Boca Raton, FL 33487-2742

and by CRC Press
4 Park Square, Milton Park, Abingdon, Oxon, OX14 4RN

CRC Press is an imprint of Taylor & Francis Group, LLC

© 2024 selection and editorial matter, Ravi Kant and Hema Gurung; individual chapters, the contributors

Library of Congress Cataloging–in–Publication Data
Names: Kant, Ravi (Professor of mechanical engineering), editor. | Gurung, Hema, editor.
Title: Industry 4.0 : concepts, processes and systems / edited by Ravi Kant and Hema Gurung.
Description: First edition. | Boca Raton : CRC Press, 2024. | Includes bibliographical references.
Identifiers: LCCN 2023010049 (print) | LCCN 2023010050 (ebook) | ISBN 9781032159492 (hardcover) | ISBN 9781032159522 (paperback) | ISBN 9781003246466 (ebook)
Subjects: LCSH: Industry 4.0.
Classification: LCC T59.6 .I355 2024 (print) | LCC T59.6 (ebook) | DDC 670--dc23/eng/20230525
LC record available at https://lccn.loc.gov/2023010049
LC ebook record available at https://lccn.loc.gov/2023010050

ISBN: 9781032159492 (hbk)
ISBN: 9781032159522 (pbk)
ISBN: 9781003246466 (ebk)

DOI: 10.1201/9781003246466

Typeset in Sabon
by Deanta Global Publishing Services, Chennai

Contents

Preface

The manufacturing sector has seen many revolutions in the past. There were majorly three revolutions which are called the first, second, and third industrial revolutions. The driving force behind these revolutions was the need to meet high demand at low cost, provide a better life to the workers and society, bring a large population above the poverty line, change the economy that was majorly based on agriculture, provide a better life to the human being, increase the wealth of society, and develop other sectors like transport, medical, construction, etc. Now, industries are going through a fourth generation revolution, popularly called Industry 4.0 or I4.0. The driving force and motives are almost similar to those which played an important role for previous industrial revolutions.

Industry 4.0 majorly emphasizes digitalization, automation, and communication. It will enable faster, flexible, efficient, and high-quality low-cost processing in the industries. Machines will be capable of interacting with the operator, designer, quality engineer, distributor, customer, and even with other machines. They will have the capability of self-learning by collecting real time data and storage from the cloud. Connectivity is one of the most important pillars of Industry 4.0 that will improve collaboration and access across departments, partners, vendors, product, and people. Industry 4.0 will offer high flexibility, self-optimization, self-configuration, self-diagnosis, cognition, and intelligent support at different levels. It will provide the advantages of quick response to change in demand, faster innovations, customized production, decentralized production, less defects, and low-cost production.

This book discusses various aspects of Industry 4.0 including the basic principles, framework, components, advantages, challenges, implementation methodology, and future aspects of this industrial revolution. The most important components of Industry 4.0 such as sensors and actuators for the smart machines, machine learning (ML), image processing, blockchain, artificial intelligence (AI), big data, reconfigurable components, additive manufacturing, and condition monitoring are discussed in this book. It will provide an overall idea of Industry 4.0 especially for the manufacturing sector. This book has 11 chapters which cover broad domains by covering literature review and original research.

Chapter 1 discusses the previous three industrial revolutions, their driving forces and components. The fourth industrial revolution (or Industry 4.0) is explained in detail with its important components and characteristics. The impact of Industry 4.0 on society and its sustainability aspects are described along with an approach for smooth transition implementation. The key technological areas related to Industry 4.0, such as the Internet of Things, autonomous robots, cyber systems, smart factories, additive manufacturing, artificial intelligence, and big data analytics, are introduced. Since seamless connectivity is the baseline for automation of machines and factories in Industry 4.0, cyber security challenges are also discussed. Aspects of preparing skilled labour, digital transformation, productivity, challenges, and economical aspects in view of Industry 4.0 are explained.

Chapter 2 represents basic concepts and applications of smart sensors and actuators for next generation digital industries. Industry 4.0 brings more and more industrial axis of motion by introducing Internet of Things (IOT) and information and communication technologies (ICT), where an interaction between digital and real world occurs through smart sensors and actuators. Smart sensors collect information from the environment and generate data corresponding to the information. Smart sensors and actuators are the backbone of the automation, intelligent manufacturing, self-monitoring, and correction.

Chapter 3 highlights the latest developments in Industry 4.0 and Industry 5.0 in various industries. The requirements, advantages, preparedness, and limitations of Industry 4.0 and 5.0 are discussed. Industry 4.0 is majorly about mass customization, but Industry 5.0 can be about using the latest technologies not only to customize but also to personalize the products and production processes. The personalized touch can deliver a better customer experience due to additional value and revenues. These personalized interactions between humans and machines can be enabled through voice command or even brain-waves. This seamlessly integrates artificial intelligence with everyday actions. Applications of Industry 4.0 and 5.0 in healthcare, transportation, infrastructure, and finance sectors are discussed. Finally, this chapter highlights legal challenges, regulatory challenges, and research directions in this sector.

Chapter 4 focuses on machine learning (ML) driven manufacturing. It covers important machine learning algorithms that help in automatizing manufacturing for high quality and low-cost production. It gives a brief overview of two of the most sophisticated machine learning techniques, support vector machines and artificial neural networks. It deals with the prevalent problems in the manufacturing industry and their impact on the modern manufacturing process and environment, and describes the ways to tackle these problems with the help of machine learning and artificial intelligence. Two of the most prevalent problems in the manufacturing industry are machinery malfunction and product quality control. This chapter describes how to predict and compensate these problems using support vector machines and artificial intelligence.

Chapter 5 discusses the application of machine learning in material science and engineering. It presents various image processing modules, libraries, and frameworks along with the problems and solutions associated with them. Subsequently, the different data ecosystems which can be useful for data collection and preparation before selecting any ML model are highlighted. Furthermore, the various opportunities and challenges that material scientists have been facing in the development of new materials are also outlined for a better understanding of ML based image processing in material science and engineering. In the end, SWOT analysis has been done using ML in material science and engineering.

Chapter 6 presents the impact of blockchain, artificial intelligence, and big data on Industry 4.0 and their challenges. Blockchain is the technology over which many applications, for example, cryptocurrency operate. The implementation of blockchain has increased in Industry 4.0. Blockchain can provide data security, transparency, traceability, decentralization, and helps in integrating data. Future machines will be smart which means they will remember the task, and will operate without an operator's commands. A large amount of data set will be used for making decisions. The management and arrangement of big data in a sequence is a big task as a large variety and volume of data keeps flowing in real-time. There are many technologies that have enhanced the productivity of the industries.

Chapter 7 presents the characteristics of enhanced sensors and improved connectivity for the successful implementation of Industry 4.0. Successful implementation majorly depends on data collection and communication between machines. Organizations need to use high-precision sensors and communicate the data collected by the sensors in real-time. Smart sensors can collect and compute the data independently, thereby reducing the quantity of communication as well as making data collection and communication possible in real-time.

Chapter 8 consists of design of modular robotic systems, modular components, robot assembly, and reconfiguration possibility. Robotic systems are an integral part of industrial automation. The concept of customized robotic manipulators is emerging due to a rapidly growing range of less repetitive applications. Industry 4.0 involves timely data collection, timely status information, history, and target states, which help in optimal planning of production systems. Demand based customized products can be planned for the best utilization of resources and services. This involves highly flexible mass production that can be rapidly adapted to market changes. Work pieces, tools machines, and robots need to be capable of autonomously exchanging information, triggering actions, and controlling each other independently. New trends toward mass customization, small scale manufacturing, and maintenance services, which are non-repetitive in nature, require robotic manipulation systems with different configurations. Modularity and reconfigurability provide cost-effective solutions for rapid customization.

Chapter 9 provides the necessary concepts for understanding the digital technologies in additive manufacturing (AM). It includes a set of technologies to obtain a physical build part through the construction of all subsequent layers by adding material. The key benefits of AM in Industry 4.0 are decreased prototyping cost, time, digitization of processes, and assembly accumulation in a single part. The role of AM in Industry 4.0 is expounded through discussing different AM technologies and their working concepts with several potential challenging applications as per future requirements.

Chapter 10 provides an overview of the recycling and utilization of metallic and polymeric waste as feedstock materials via AM technology. AM technologies are becoming popular in various sectors like automobiles, aerospace, lab prototyping models, architectural models, printing electronics, and the construction industry. Despite the recent research and developments in AM technology, there is still demand for improvements in feedstock materials and methods to match the traditional manufacturing technologies for mass production. Handling industrial waste, such as metal scrap and packing materials, is challenging for efficient production and environmental protection. Thus, the recycling of industrial waste as potential feedstock material for AM technology to produce useful products is highly desirable, and is covered in this chapter.

Chapter 11 discusses tool condition monitoring during mechanical micromachining. Tool condition monitoring using sensors has become an integral part of micro machine tools to enhance product quality and productivity. The breakage of the tool causes poor surface quality and dimensional accuracy for machined parts, or possible damage to a workpiece or machine.

The chapters covered in this book introduce the readers to the basic concepts, components, pillars, tools, and applications of Industry 4.0. They educate the readers with the knowledge required for implementation, functioning, operation, improvement, and repair and maintenance of Industry 4.0 enabled smart machines or systems. Each chapter covers a good amount of literature review from recent research. The chapters are a good combination of review articles and original research works. They highlight the research gaps, directions for future work, and path to the upcoming fifth industrial revolution. The content of this book will be useful for academicians, researchers, and practicing engineers who are working on Industry 4.0.

Editors: Ravi Kant and Hema Gurung

Editors

Ravi Kant is an Assistant Professor in the Department of Mechanical Engineering at the Indian Institute of Technology Ropar, India. He earned his Bachelor degree in Mechanical Engineering from Maharshi Dayanand University, Rohtak (Haryana, India). He completed his M. Tech. from the Department of Mechanical Engineering at the Indian Institute of Technology (IIT) Guwahati, India with specialization in Computer Assisted Manufacturing. He worked on the Investigation on Formability of Adhesively Bonded Sheets during his M. Tech. project. He also earned his doctorate from IIT Guwahati in the field of laser forming process. His research interests include laser transmission welding, hybrid machining, laser forming, cold spray coatings, additive manufacturing, hybrid joining, and sustainable materials. He has completed many research projects and consultancy works in these research areas. He has contributed about 100 research articles in peer-reviewed journals, conferences, and edited books. He has edited three books in the field of design, manufacturing and materials. He has also guest edited five special issues in reputed journals. He has developed and taught advanced courses such as modern manufacturing processes; sustainability science and technology; analysis of casting, forming and joining processes; advanced welding technology; micromanufacturing; manufacturing, among others. He has conducted various international conferences, workshops, symposiums, colloquiums, and faculty development programs in the field of advanced manufacturing technology.

Hema Gurung is an independent researcher in the field of Robotics. She earned her bachelor degree from Kalyani University (West Bengal, India) in Electronics and Instrumentation Engineering. She completed her M. Tech. in Mechatronics from IIEST (Indian Institute of Engineering and Science Technology, Shibpur, India) and earned her Ph.D. from the Department of Mechanical Engineering at the Indian Institute of Technology Guwahati, India in the field of Robotics. She has worked in Hanbat National University, South Korea and Thapar University, India. Her research work includes experimental analysis, numerical simulation, optimization, control systems, state estimator, robotics, smart sensors and actuators. She has published several research articles in international journals, conferences, and edited books. Currently she is working on the application of smart sensors and actuators in Industry 4.0.

Contributors

Baburaj M
Indian Institute of Technology
    Tirupati
Andhra Pradesh, India

Niraj Bala
National Institute of Technical
    Teachers Training and Research
Chandigarh, India

Shilpi Chaudhary
Punjab Engineering College
Chandigarh, India

Arvind Kumar Gupta
Indian Institute of Technology
    Ropar
Punjab, India

Hema Gurung
Independent Researcher, India

Ravi Kant
Indian Institute of Technology
    Ropar
Punjab, India

Arun Kumar
Indian Institute of Technology
    Ropar
Punjab, India

Manoj Kumar M
Muthoot Institute of Technology
    and Science
Kerala, India

Rishabh Machhan
Punjab Engineering College
Chandigarh, India

Dr. Ashok G Matani
Retired Professor
Government College of Engineering
Jalgaon, Maharashtra, India.

Nithin Tom Mathew
BITS Pilani, Rajasthan, India

Naman Krishna Pande
Indian Institute of Technology
    Ropar
Punjab, India

Dipak A. Patil
Indian Institute of Technology
    Ropar
Punjab, India

Ayush Pratap
Indian Institute of Technology
    Ropar
Punjab, India

Sumitkumar Rathor
Indian Institute of Technology
    Ropar
Punjab, India

Narpat Ram Sangwa
Indian Institute of Management
    Sirmaur
Himachal Pradesh, India

Kuldip Singh Sangwan
Birla Institute of Technology and
    Science Pilani
Rajasthan, India

Neha Sardana
Indian Institute of Technology
    Ropar
Punjab, India

Harpreet Singh
Indian Institute of Technology
    Ropar
Punjab, India

Malkeet Singh
Indian Institute of Technology
    Ropar
Punjab, India

Perminderjit Singh
Punjab Engineering College
Chandigarh, India

Sandeep Singh
Punjabi University

Ekta Singla
Indian Institute of Technology
    Ropar
Punjab, India

Rupen Trehan
Punjab Engineering College
Chandigarh, India
Punjab, India

Chapter 1

Industry 4.0

Its evolution and future prospects

Sandeep Singh and Niraj Bala

1.1 INTRODUCTION

"During the last few years, computers will be integrated into mostly industrial products" (Karl Steinbuch, 1966). Global rivalry in the field of the manufacturing industry enhances day by day. Furthermore, the United States and the German industry have observed the progress of the Internet of Things and its services in the field of manufacturing engineering. The United States is taking measures to address deindustrialization via programs to enhance advanced manufacturing [1]. The industry-based fairs, funded projects, and conferences are almost the integrated part of Industry 4.0 in the regions of Austria and Germany. The generally considered fourth generation of the industrial revolution implies this is within reach. However, what is the exact meaning of industry 4.0?

1.1.1 Industry evolution 4.0

The integration of internet technology into the industry is the central concept of Industry 4.0. Industrial production is currently experiencing major issues in communication technologies such as Embedded Systems (ES) and the Internet of Things (IoT). Industry 4.0 is an integrative word, a perspective that illustrates where the path/experience in manufacturing and industrial production is leading. Many organizations are already inadvertently on this journey by implementing certain aspects of Industry 4.0 principles nowadays.

Figure 1.1 depicts that during the end era of the 18th century, the beginning of industrialization with the advent of mechanical equipment used in production such as mechanical tools for parts manufacturing took place [2]. The first industrial rebellion was accompanied by the second towards the late 19th century with the introduction of electrically driven machinery employed for mass manufacturing depending on the specialization of tasks. In the beginning of the 1970s, the third industrial transformation took place. It was premised on the automation of industrial processes via the use of information and electronics technology. Widespread computing

DOI: 10.1201/9781003246466-1

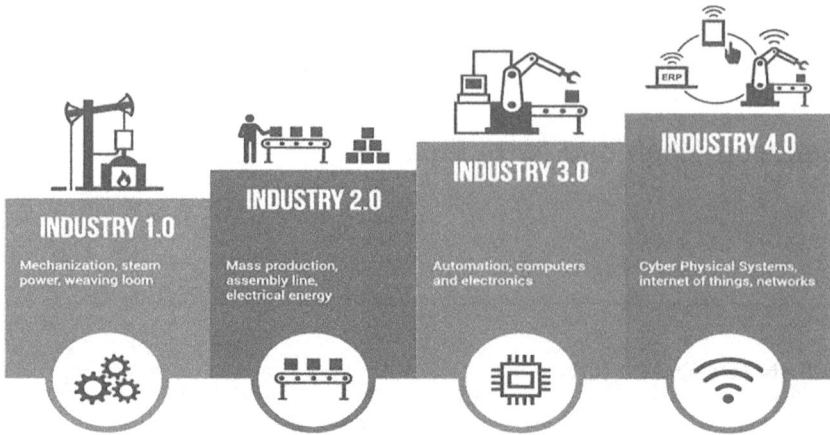

Figure 1.1 Stages of industrial revolution [2]

has emerged as a reality as a consequence of consistent development and the inevitable growth of the Internet. The embedded systems are rapidly being linked to one another and the Internet [3]. As a result, the real and the digital world merge to form the cyber-physical systems that are the accelerators of the fourth industrial transformation. The terminology Industry 4.0 (also known as an integrated industry in the United States) has become common. To facilitate Industry 4.0 in the future, two key techniques are proposed: Physical processes and computing integrated with cyber-physical systems [4]. Networks monitor integrated computers and regulate physical processes, typically through feedback loops which influence calculations and vice versa. Cyber-physical networks are the next evolutionary step beyond currently integrated systems, serving as the framework for an Internet of Things that, when combined with the services of the Internet, enables Industry 4.0 [5, 6]. Internet technology – the next stage of the Internet – is a worldwide system of computer networks, gadgets, sensors, and machines. By combining the real world with the digital world of software and the Internet, businesses and consumers may build and experience new services based on web-based/internet business strategies. This will have a significant influence on how we do business [7, 8].

1.1.2 Smart factory

Smart goods, procedures, and activities are the heart of Industry 4.0. The smart factory, as depicted in Figure 1.2, is thus a part of Industry 4.0 [9]. The advanced factory helps to support the rapidly increasing intricacy and effectiveness of manufacturing. There exists a direct interaction between people, manufacturing facilities, machines, and transporting and storage systems in the smart factory. Smart goods are aware of their production process and potential uses. They actively assist the manufacturing process

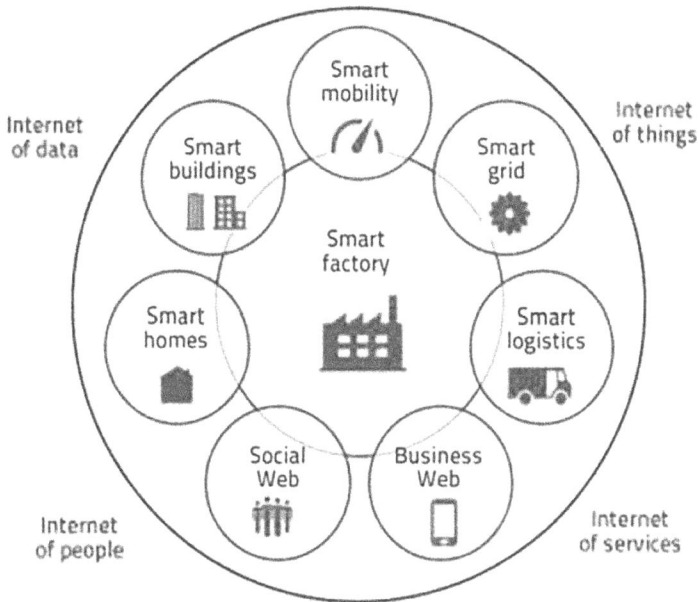

Figure 1.2 Internet of data and internet of services are a part of smart factory and Industry 4.0 [9]

and documentation with this knowledge ("when was I created, where should I be forwarded to and which characteristics should be utilized to generate me?"). Industry 4.0 signifies a paradigm shift from "centralized" to "decentralized" production enabled by technological advancements, resulting in an inversion of conventional manufacturing process logic [10].

This implies that industrial manufacturing technology no longer merely produces the product, but instead interacts with the machinery to instruct it specifically on what to do. With its interconnection to smart systems, smart grids, smart logistics, and smart buildings, the smart factory is a key component of emerging smart structures. Traditional value chains will be improved, and new industry models will emerge [11, 12]. Previously, industrial revolutions were characterized by increasing affluence. They have also generally had far-reaching social and climatic impacts, both favorable and detrimental. The upcoming section investigates the social and environmental consequences of Industry 4.0.

1.2 SUSTAINABILITY CONSEQUENCES

Primarily, the goal of Industry 4.0 is to increase and develop a company's long-term viability by boosting production flexibility and effectiveness through interaction, data, and knowledge. The three upstream stages of the fourth industrial revolution have an impact both within the company (such as production parameters, innovation, association, and staff members) and outside the company (e.g., on the society and environment) [13]. An elaborated examination of the literature reveals that the influence on future work, as well as the consequences on the environment, are discussed widely.

1.2.1 Future work

Various researchers concur that despite more automation, people's power will continue to play a vital role in Industry 4.0. In the scientific research on future production work, 60.2 percent of participants believed that human employment will continue to play a significant role in future production. Furthermore, 36.6 percent believed that human tasks had an essential influence. There will still be typical manually operating operations in Industry 4.0 that require suitable human talents such as knowledge, innovation, awareness, or adaptability [14, 15]. However, people's positions within the organization and the duties they have to accomplish will alter. Beier et al. mentioned this modification as follows, "In future manufacturing, there will be more coordinators and factory supervisors. Machines provide hard muscular effort as well as some mental work" [16]. The powerful network of people and machines within Industry 4.0 necessitates considerable changes in work procedures, work environment, job content, and necessary skills. As a result, new occupations for employees will arise [17, 18]. As physical jobs become less significant in the future, knowledge in the area of new information technologies, such as programming, planning, execution, and error correction, will emerge progressively essential [19–21]. Furthermore, the requirement for integrated knowledge and abilities to recognize how comparable fields and faculties function and think becomes more prominent.

The physical burden on individuals will decrease as the amount of mechanization and communication among people and machines increases. This is especially helpful because the proportion of older workers will rise as a result of the age shift [22]. Therefore, job prospects, particularly for old people, are increasing. Greater psychological stress is predicted in conjunction with decreased physical requirements. On the one hand, this is due to the ongoing change in job content, which necessitates more flexibility and response [23]. On the other hand, it causes psychological stress if interaction and cooperation among employees declines because of enhanced communication between computers and humans, or if task distribution no longer comes from a supervisor but from an information system [24, 25]. Nowadays, it is clear that the mental trauma is increasing day by day. Industry 4.0 might aggravate this [26]. As a result, it is essential to incorporate this risk into the development of Industry 4.0.

Although more flexibility may result in increased psychological stress, it may also have a good impact on employees' work-life balance. As a result, the work structure may be better fitted to the demands of employees in terms of balancing work and personal life. Personal and professional growth can also be integrated more effectively [27, 28]. This advantage may be especially important for SMEs in the future to combat a skilled labor shortage. Because of the demographic situation, empirical studies indicate a scarcity of competent workers in German-speaking nations [29]. Finding skilled people and retaining them, in the long run, is getting increasingly challenging, particularly for SMEs. Thus, the benefits of Industry 4.0 mentioned above might give a competitive advantage in the struggle for workers.

In conclusion, it should be highlighted that quantitative assertions regarding the influences of Industry 4.0 on the number of workers are difficult to make at this time. However, it is expected that basic manual duties will decrease. A suitable professional growth toward Industry 4.0 must be addressed for the impacted group of employees in due course [30, 31]. Otherwise, this might lead to socio-economic issues. From a qualitative standpoint, it is clear that the occupations and abilities of production staff will change. At this stage, educational institutions, in addition to industrial enterprises, are being asked to consider future training requirements [32].

1.2.2 Ecological feedback

Production has several detrimental consequences on the environment. Besides, to the intended value some unfavorable side effects (e.g., emissions, natural resource use, and energy consumption) are produced, which might cause environmental harm. Several initiatives have been made in Europe over the years to mitigate these detrimental impacts, which pose problems to the industry. Compliance with high environmental standards, today's and prospective emission exchanging programs, CO_2 reduction objectives, or raw material and energy price patterns are just a few examples of factors that are progressively affecting industrial firms' competitiveness [33]. Huge utilization of raw materials in industrial production has the outcome not only of more expenses but also of enhancing environmental consequences. A major goal shows resource efficiency, which is primarily characterized by the company's production [34, 35]. As a result, energy and resource-efficient systems will become progressively essential.

Industry 4.0 can help by providing continual energy and resource monitoring. By giving complete information on each stage of the manufacturing process, resource and energy consumption may be minimized across the whole value supply chain. Furthermore, systems may be continually improved during the manufacturing process in terms of resource and energy usage, as well as emission output. This can provide a prominent subsidy to the company's long-term success [36, 37]. It is also feasible to address resource and energy efficiency throughout the company's development stage by optimizing rooms, areas, passageways, or lines, designing decentralized supply and building energy cycles [38].

The CO_2 emissions can be reduced by optimizing the industrial manufacturing process. These reduced pollution limits make it easier to establish factories in urban areas. In addition to social advantages (improved work-life balance), such urban production brings environmental benefits (e.g., reduced emissions from daily transportation to work) [39].

The eventual disposal of items or equipment will also be facilitated by more documentation within Industry 4.0. Even the most complicated technical equipment may be deconstructed into its constituent parts at a reasonable cost and then disposed of or recycled using the knowledge obtained. Material cycles can be closed and optimized using technologically complicated recovery procedures [40]. This offers an efficient cost reduction along with an effective commitment to resource stewardship. Future production is frequently considered in conjunction with new industrial technology. According to experts, additive manufacturing techniques such as 3D printing will play a significant role. This manufacturing method allows for the creation of products with precise qualities that are created layer by layer [41, 42]. 3D printing employs an enormous amount of materials, including gypsum, ceramics, polymers, and metals. 3D printing is ideal for producing one-of-a-kind items or small series (e.g., prototypes or highly specialized products). Because fewer resources are required, waste is rarely created, and production and transportation logistics may be minimized through decentralization; 3D printing is regarded as an ecologically favorable production technique. [43–45].

The next section focuses on the state of digital production to apply theoretical features realistically (the premise for Industry 4.0). This will be done in light of an empirical study conducted in Austria.

1.3 EMPIRICAL ANALYSIS

What does Industrial Revolution 4.0 imply for the software industry? Vertical IT accommodate refers to the accommodation of numerous IT-Systems at various administrative levels (such as on the shop floor, there are actuators and sensors, in production management, there are manufacturing execution systems, and on the corporate planning level, there is an ERP-System) in the areas of manufacturing and automation [46]. Industry 4.0 is not feasible without high-quality master and transaction data, well-defined IT-based processes, and the capacity to do complicated data analysis [47]. The following comments investigate the situation of digital production in Austria, which is a requirement for Industry 4.0. The Kapfenberg/Austria Institute of Industrial Management conducted a study on the state of digital production with a focus on SMEs. Of the 4,725 distributed survey forms to SMEs and the 222 returned replies, 136 records were considered for assessment. Figure 1.3 shows that digital production is currently under-utilized in Austrian small and medium-sized businesses [48].

ERP and MES systems are commonly utilized in large corporations; however, they are infrequently employed in small and medium-sized businesses (ERP 46 percent, MES seven percent). It should be highlighted that in SMEs, there is a scarcity of software support and vertical IT-Integration, and paper-based data interchange predominates [49].

As the number of Austrian MES suppliers is limited, the second portion of the study concentrated on MES providers from Austria and Germany. Throughout, all future developments were seen as significant. According to MES providers polled, there is a strong trend toward "green production" and "work integration of modified machines in an established IT landscape" (Figure 1.4) [48].

Besides these basic findings, the study also revealed conclusions on sustainability. The human being's importance in Industry 4.0 has already been discussed extensively. This is also corroborated by the findings of the study. A significant proportion of respondents see people as the most important successful element to accomplish the adoption of IT systems. This conclusion may be applied to Industry 4.0 as well, because workers' technical knowledge and the board leadership qualities, as well as the project team's makeup, are important to success [50–52].

According to the report, green manufacturing in general and resource and energy management are top priorities for nearly 45 percent of the examined companies. As a result, the deployment of energy and resource management as part of Industry 4.0 is often seen as favorable [53]. Aside from the openness of the energy intake, the possibility of allocating energy prices to goods, process stages, or departments is seen favorably. Furthermore, the

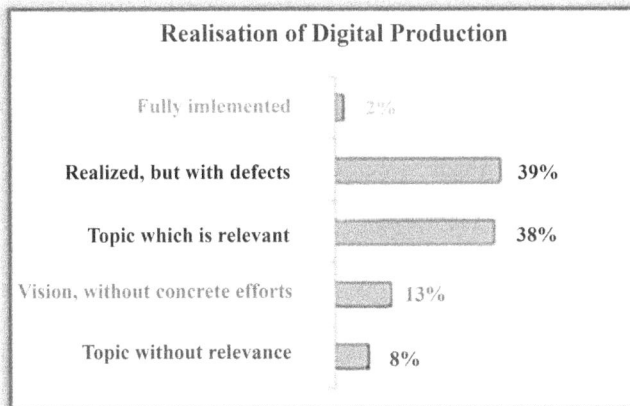

Realisation of Digital Production

Fully imlemented	2%
Realized, but with defects	39%
Topic which is relevant	38%
Vision, without concrete efforts	13%
Topic without relevance	8%

Figure 1.3 Medium and small scale industries' digital production in Austrian [48]

IT Trends in Production

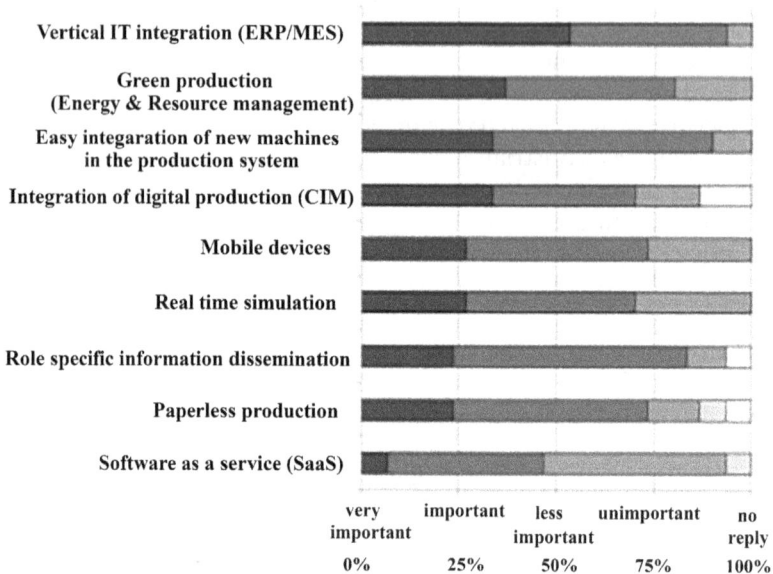

Vertical IT integration (ERP/MES)				
Green production (Energy & Resource management)				
Easy integaration of new machines in the production system				
Integration of digital production (CIM)				
Mobile devices				
Real time simulation				
Role specific information dissemination				
Paperless production				
Software as a service (SaaS)				

very important	important	less important	unimportant	no reply
0%	25%	50%	75%	100%

Figure 1.4 Digital production future trends [48]

firms surveyed feel that the subject of green production will become more significant in the future [54].

A global study performed by CSC-Computer Services also reveals identical alarming results. More than half of the executives polled in Switzerland, Austria, and Germany have never heard of Industry 4.0 [55]. Only a quarter of the population has a thorough understanding of the changes brought on by Industry 4.0. Despite this, just 13 percent of Austrian businesses polled plan to undertake training or education programs in this area. Eighty-four percent of respondents believe they are under-informed about Industry 4.0's threats and potential. Industry 4.0 is predicted to enhance efficiency by 50 percent, cost reduction by 43 percent, productivity by 40 percent, customer satisfaction by 40 percent, and competitiveness by 39 percent.

1.4 BENEFITS OF INDUSTRY 4.0 RELATED TO OVERALL CONTINUAL DEVELOPMENT

Industry 4.0 benefits include increased profitability, increased flexibility and agility, and improved production and efficiency. Industry 4.0 also enhances the experience of customers. While advanced factory characteristics are fascinating, discussions about them should always focus on the advantages of

Figure 1.5 Benefits of Industry 4.0

Industry 4.0. Any investment made in new technology, enhanced manufac-turing techniques, or enhanced systems needs to yield a return. Industry 4.0 technologies have several advantages, thus there is good ROI potential [56]. This includes systems for automation, managing production, and making decisions. The most significant benefits of Industry 4.0 in several industries are listed below and are also depicted in Figure 1.5.

(a) **Improved productivity**

   Industry 4.0 technology often enables doing more with less. It aids in increasing output and speed while allocating resources more eco-nomically and effectively [57]. Overall equipment effectiveness will increase as we come closer to being an Industry 4.0 smart factory.

(b) **Enhanced efficiency**

   Numerous components of the manufacturing process will become accurate as a result of Industry 4.0. Effective advantages like less machine downtime and the potential to create more products more quickly have already been addressed. Among the many examples of improved efficiency are quicker batch changes and automatic report-ing. Other business decisions, such as new product introductions, become effective [58].

(c) **Improved sharing of knowledge and teamwork**

   In the conventional sense, manufacturing facilities operate as com-partments. Silos are distinct machinery within a facility, much like distinct facilities. Teamwork and information sharing consequently suffer. Your manufacturing lines and departments can communicate with one another using Industry 4.0 technologies, regardless of place and platforms [59]. This makes it possible for an organization to share information obtained, for example, by a sensor on a machine in one factory. The finest part is that it can be done automatically, i.e., system to system and machine to machine, without requiring human contact. The data from a single sensor might be used in real-time global manufacturing line optimization.

(d) **Flexibility**

One benefit of Industry 4.0 is increased agility and adaptability. For instance, scaling up and down output is simpler in a smart factory. Adding new products to the production line is also made simpler, and one-off manufacturing runs and other alternatives are made possible [60].

(e) **Conformity**

Compliance with legal requirements should not be a time-consuming process in industries like the production of pharmaceuticals and medical devices [61]. Industry 4.0 enables the automation of compliance activities such as tracing, quality, serialization, data logging, etc.

(f) **Good experience of customer**

Additionally, Industry 4.0 offers opportunities to enhance customer support and the shopping experience. Thanks to its automatic track and trace features, problems can be quickly resolved. Additionally, there will be less issues with availability of a product, and its quality will increase [62].

(g) **Research opportunities**

You may better comprehend your supply chains, distribution systems, and even the products you produced thanks to Industry 4.0 technology [63]. This brings up possibilities for innovation, including, among other things, altering a company procedure, creating a unique product, and optimizing a supply chain.

(h) **Enhanced profitability**

This Industry 4.0 benefit will be attained in part thanks to many of the previously mentioned factors, including higher revenues and lower expenses. Additionally, Industry 4.0 technologies make it possible to produce goods with larger margins, more distinctive designs, or both. For instance, Industry 4.0 technologies allow companies to continue using mass manufacturing methods while still offering customers individualized products [64].

(i) **More revenue**

By automating the manufacturing line and integrating other industry technology, Industry 4.0 can introduce a shift with low costs of staffing to meet an increase in demand for a new contract [65].

(j) **Investment return**

Industry 4.0 is revolutionizing manufacturing all around the world. The Industry 4.0 benefits and the possible investment return are the most important [66]. If you want to maintain your competitiveness and outfit the time to consider the next phase of your 4.0 journey is right now.

1.5 HOW INDUSTRY 4.0 WILL CHANGE THE SOCIETY

The industrial 4.0 revolution will be characterized by the digitalization of business and the computerization of production. Big data analysis, machine learning, manufacturing, and transportation are only a few examples of the increasingly efficient use of intelligent and autonomous systems. The most typical application of Industry 4.0 will be smart factories. However, it might result in intelligent ports with completely automated cranes that can load various materials onto ships without assistance from a person [67]. The transportation industry will be significantly impacted by the rising automation of cars, trains, and other forms of transportation that will reduce the need for drivers and pilots.

(a) **Analysis of big data:** Analytics that can examine and collect appropriate insights from large amounts of data are known as analytics of data. Big data analytics is getting better all the time, and it will play a bigger role in the scenario of Industry 4.0 [68].

(b) **The use of 3D technology:** 3D technology is a cutting-edge field that is currently being promoted by businesses to improve the shopping experience and to clarify the working process, including 3D printing, 3D visualization, and 3D modeling [69].

(c) **Smart factories:** These facilities use cutting-edge technology like big data processing, cloud computing, robots, strong cyber-security, and smart sensors to operate in a highly efficient, secure, and economical manner [70].

(d) **IoT platforms:** The term "Internet of Things" (IoT) refers to any device that can collect data and move it on the Internet, and exchange signals with other devices. Modern toasters, lighting, and refrigerators are examples of IoT [71].

(e) **Location monitoring technologies:** These techniques, which provide location tracking, are frequently encountered on mobile devices. It is simpler to communicate whereabouts with reliable people while using this technology [72].

(f) **Sophisticated algorithms:** Machine learning algorithms are incredibly complex mathematical formulas that provide guidelines for computers to carry out various tasks [73].

(g) **Immersive technology:** Augmented reality refers to wearable devices like Google Glass that add visual information to reality [74].

(h) **Advanced sensors:** Advanced sensors can gather data then process it, and display digital data when the time is right [75].

(I) **Data representation:** Typically, data is expressed visually using layouts, graphs, and slideshows. In the future, it will be possible to see larger volumes of data and represent it in a variety of ways [76].

(j) **Fraud detection and prevention:** Industry 4.0 uses big data and pattern recognition to notify parties when fraud is present [77]. Banks and other financial institutions will depend more and more on this technology in the future.

(k) **Network technology:** The well-known digital currency Bitcoin gave rise to this technology. On global public ledgers, data is stored as "blocks" that are connected by a "chain" and verified by "miners" [78].

(l) **Profile of customer and multilevel engagement of customer interaction:** Based on identifiers, this technique classifies clients. Identifiers include things like interests, geography, and hobbies [79].

(m) **Cloud technology:** Data storage that is not dependent on local servers, PCs, or laptops is known as cloud technology. It is dependent on "the cloud," which is a networked memory system that is located far away [80].

(n) **Advanced human-machine interface:** The advanced interface takes place when a machine shows operators visual data about the operations it is carrying out in real time [81]. This enables them to understand how the machine is operating.

(o) **Accessories:** The various devices that individuals use to access the Internet on the go, such as iPads, e-readers, Apple watches, and PCs. These gadgets, which will be widely used, will be essential for Industry 4.0 [82].

1.6   THE IMPACTS: USE OF TODAY'S TECHNOLOGY HAS GREATLY INCREASED PRODUCTIVITY, PERSONALIZATION, AND AUTOMATION

Like all past industrialization, digitalization will significantly advance humankind's capacity to produce goods, transmit them globally, and optimize customer engagement. Industry 4.0, in contrast to the earlier industrial era, enables historically unbelievable automation, customization, and profitability [83]. Additionally, the enormous amount of development presently occurring in several sectors, including materials science, microbiology, fuel cells, particle physics, 3D nanotechnology, self-driving, and storage systems, should progress significantly and effectively.

In the upcoming years, a lot of people will use driverless vehicles, tissue engineering will revolutionize medicine, space exploration will be improved and expanded, 3D printers will modify how goods are created and delivered, and many other significant developments will occur [84]. Even though there are many positive advancements that Industry 4.0 is projected to bring to humanity, this timeframe will also offer a significant number of challenges.

1.7 NEW PROSPECTS AND CHALLENGES

The transformation of employment opportunities brought on by digitization, improved robotic technology, and machine learning will probably be Industry 4.0's biggest issue. The ability of intelligent machines to perform ever more complicated tasks may put thousands of jobs at risk. Automation might displace up to 40 percent of present jobs within the next 15 years [85].

As a consequence, there will be unemployment in the years ahead. However, this is a common occurrence during industrialization. For instance, the invention of the tractor in the past caused thousands of agricultural workers to lose their jobs. Emerging technological revolutions produce a huge volume of modern jobs. Digitalization will therefore both create and diminish opportunities [86].

Just as prior decades would have to adapt to technologies 1.0, 2.0, and 3.0, so too people and enterprises must shift to Industry 4.0. All of the amazing new technology will be available to anyone who can adopt it rapidly [87]. Ultimately, digitalization has the power to significantly enhance the lives of millions worldwide. There will be huge improvements in just about everything, including job stability and Internet commerce [88].

1.8 HOW INDUSTRY 4.0 IMPROVES THE PRODUCT DESIGN

The use of networked and smart technologies is the foundation of the Industry 4.0 approach, which is progressively changing how manufacturing processes are carried out. Digital innovations from the fourth manufacturing generation have an influence on economic equipment design and engineering as well as how businesses generate commodities. Technology 4.0 enables major improvements in quality design by incorporating subprograms that are connected to machinery or other tangible objects that allow the acquisition of data collection. At every phase of the product development cycle, especially design, the Internet of Things may be used to create intelligent things [89]. Engineers may maintain a close check on a product's effectiveness and usage by possessing access to records and statistics when it is in service. For instance, authentic surveillance enables the network of digitized products to be effectively redesigned to save considerable amounts of energy, decrease interruption, and improve production processes. IoT is being used by larger companies in the automotive sector to handle various safety procedures for their automobiles. Cars featuring stop-and-go support, three-dimensional maps, or other features, for instance, make it possible to gather data that engineers may use to make driving better for customers [90]. Production companies will realize that they need to depend on information technologies to improve both the production of commercial products and their development. The acceptance of 4.0 must be coupled with the execution of management processes that permit the storage, communication, and utilization of information that interacts with the product to get the advantages of intelligent manufacturing [91].

1.9 CHALLENGES AND RESEARCH DIRECTION

Industry 4.0 implementation generates a vast scope of technological challenges with substantial consequences for several present manufacturing engineering characteristics. Therefore, it is crucial to establish a framework for all organizations associated in the entire supply chain and come to a consensus on safety concerns and the proper facilities before progress begins [92]. A lot of authors also assert that implementing Industry 4.0 is a challenging endeavor that will require several years to achieve. There are many factors to take into account, such as obstacles to overcome, and economic, budgetary, and political concerns while using this highly innovative production process [93, 94]. Figure 1.6 illustrates the various difficulties in processing, visualization, and manufacturing methods.

The difficult components for businesses intending to adopt the approach are the individual qualifications and capabilities of the employees, such as problem competencies, faulty identification, and the ability to engage with fast modifications and completely new duties. They ought to be permitted to test out certain Industry 4.0 techniques that demand higher degrees

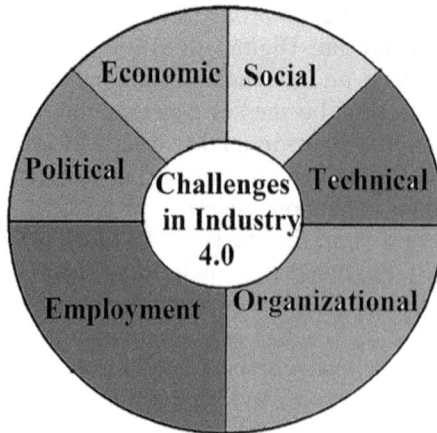

Figure 1.6 Challenges in Industry 4.0

of complexity, like data collection, analysis, and visualization throughout manufacturing [95, 96]. The formation of Industry 4.0 can substantially alter a variety of domains beyond the manufacturing area.

Other business challenges and problems incorporate creativity, technical elements, advancements in digitalization, and rising connectivity trends, all of which are important in any organization. Manufacturing 4.0 is closely related to the complete automation of all real facilities and the integration of all network stakeholders into the virtual domain. Micro and intermediate enterprises are reluctant to start the digital conversion process, and this reluctance is caused by a variety of interrelated challenges faced by factories that have achieved minimal to no success in Industry 4.0 [97].

Iyer (2018) asserts that the present state of production presents both opportunities and challenges; therefore, neither corporation administrators nor officials can build on past solutions in the modern production environment [98]. As a consequence, companies will have a lot of trouble resolving decision-making based on several factors, such as salaries, stocks, demands, transportation, and so forth. A connected company that can react rapidly and easily to changing situations while simultaneously seeking long-term opportunities with the aid of "big information" and analytics may emerge as a new form of global industrial business. The biggest challenges are still gathering relevant data and assessing the potential effects since we possess all the necessary factors to make Industry 4.0 a reality, including Internet technology, affordable IoT devices, and a specified interface framework [99].

1.10 STRATEGY FOR EMPLOYMENT OF INDUSTRY 4.0

Organizations embrace Industry 4.0 projects progressively, and there is no single set of standards for implementing changes. The initiatives which have already finished include everything from single devices to entire industrial lines. Upcoming industries will be intelligent workplaces, thanks to these cutting-edge production lines. Along the approach to Industry 4.0, production is being evaluated from traditional production to automation technology. There are organizations, such as the steel industry, that have recognized kinds of changes because they emphasize a conventional method of production [101, 102]. Industries like the automotive and home appliance sectors, which are representative of Industry 4.0, are using an expanding series of innovative types of changes. Industry 4.0's implementation procedure is shown in Figure 1.7.

According to the research analysis, the steps that can be used to help enterprises adopt Industry 4.0 include:

1. The two types of development, modern and traditional;
2. The smart industry is more substantial and advanced than the conventional industry.

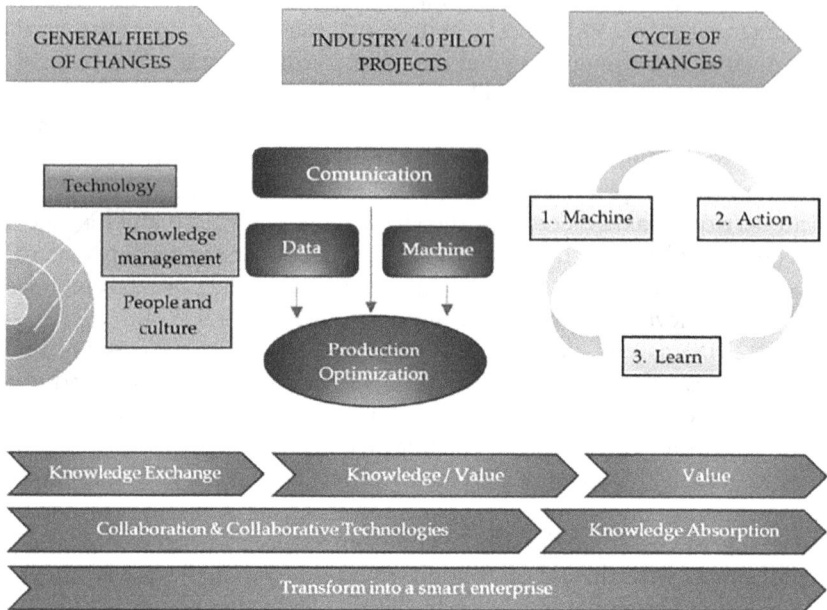

Figure 1.7 Employment of Industry 4.0 [102]

Figure 1.7, which shows the overarching paradigm for adopting Industry 4.0, provides a summary of the subject. The key regions of transformation which will result in the development of a digital workplace are represented by this framework. Smart manufacturing plants add capabilities for businesses and markets. Personalized products are replacing surplus ones in the marketplace. Customers can also receive services more quickly, easily, and conveniently [103].

1.11 CONCLUSION

Industry 4.0 is a major shift toward decentralized and customized manufacturing that will allow new business models and Internet-based services. The Internet and networking are key components of Industry 4.0. Conventional supply chains will be replaced by highly adaptable supply networks. In such value-added networks, small and medium-sized businesses will play a crucial role. Industry 4.0 has the potential to provide a solution to future difficulties. The industrial scene in Austria is characterized by a significant number of businesses that frequently create highly creative goods in an international scenario.

The findings of the literature research and empirical study reveal that Industry 4.0 provides organizations with several possibilities. Increased automation can give a competitive edge, especially in a high-wage nation like Austria. Simultaneously, environmental issues can be easily resolved. Industry 4.0 comprises USPs in worldwide competitiveness, particularly for SMEs, such as lot size one, fast response to customers, excellent quality, and flexibility. Impacts on the labor economy are also varied. Industry 4.0 might be viewed as a possibility in the competition for qualified professionals, besides the disadvantages already highlighted.

Some industrial organizations have already taken the initial steps toward Industry 4.0, but the journey to Industry 4.0 is an evolutionary rather than a revolutionary one. Technology in the areas of information, automation, and manufacturing will be more linked than ever before. For Industry 4.0, networking isn't just a goal; it's a must. Continuous IT networking, on the other hand, is almost non-existent in Austrian small and medium-sized businesses, even though it is a requirement for Industry 4.0. Its approaches currently exist; however, full adoption will most likely take a long period. Experts differ in estimating a period of up to 20 years.

REFERENCES

1. Van Neuss, L., 2018. Globalization and deindustrialization in advanced countries. *Structural Change and Economic Dynamics*, *45*, pp. 49–63.
2. Pisal, R., Razdan, S. and Kalaskar, P., 2019, December. Influence of industry 4.0 on manufacturing processes *IEEE Pune Section International Conference (PuneCon), MIT World Peace University, Pune, India,* INSPEC Acession No: 19669813, (pp. 1–5). DOI: 10.1109/PuneCon46936.2019.9105836.
3. Cabanes, Q., Senouci, B. and Ramdane-Cherif, A., 2021. Embedded deep learning prototyping approach for cyber-physical systems: Smart LIDAR case study. *Journal of Sensor and Actuator Networks*, *10*(1), p. 18.
4. Oztemel, E. and Gursev, S., 2020. Literature review of industry 4.0 and related technologies. *Journal of Intelligent Manufacturing*, *31*(1), pp. 127–182.
5. Zhong, R.Y., Xu, X., Klotz, E. and Newman, S.T., 2017. Intelligent manufacturing in the context of industry 4.0: A review. *Engineering*, *3*(5), pp. 616–630.
6. Shaji, R.S., Dev, V.S. and Brindha, T., 2019. A methodological review on attack and defense strategies in cyber warfare. *Wireless Networks*, *25*(6), pp. 3323–3334.
7. Kumar, S., Tiwari, P. and Zymbler, M., 2019. Internet of things is a revolutionary approach for future technology enhancement: A review. *Journal of Big Data*, *6*(1), pp. 1–21.
8. Sarowar, M.G., Kamal, M.S. and Dey, N., 2019. Internet of things and its impacts in computing intelligence: A comprehensive review–IoT application for big data. In *Big Data Analytics for Smart and Connected Cities*, ISBN: 9781522562078,(pp. 103–136), DOI: 10.4018/978-1-5225-6207-8
9. Bekić, D., Halkijević, I., Gilja, G., Lončar, G., Potočki, K. and Carević, D., 2019. Examples of trends in water management systems under influence of modern technologies. *Građevinar*, *71*(10), pp. 833–842.

10. Ang, J.H., Goh, C., Saldivar, A.A.F. and Li, Y., 2017. Energy-efficient through-life smart design, manufacturing and operation of ships in an industry 4.0 environment. *Energies, 10(5)*, p. 610.
11. Kolokotsa, D., 2016. The role of smart grids in the building sector. *Energy and Buildings, 116*, pp. 703–708.
12. Mohanty, S.P., Choppali, U. and Kougianos, E., 2016. Everything you wanted to know about smart cities: The internet of things is the backbone. *IEEE Consumer Electronics Magazine, 5(3)*, pp. 60–70.
13. Silvestre, B.S., 2015. Sustainable supply chain management in emerging economies: Environmental turbulence, institutional voids and sustainability trajectories. *International Journal of Production Economics, 167*, pp. 156–169.
14. Lee, X., Yang, B. and Li, W., 2017. The influence factors of job satisfaction and its relationship with turnover intention: Taking early-career employees as an example. *Anales de Psicología/Annals of Psychology, 33(3)*, pp. 697–707.
15. Fernandez, S. and Rainey, H.G., 2017. Managing successful organizational change in the public sector. In Robert F. Durant and Jennifer R.S. Durant, *Debating Public Administration.* . Routledge, ISBN: 9781315095097, pp. 7–26).
16. Beier, G., Ullrich, A., Niehoff, S., Reißig, M. and Habich, M., 2020. Industry 4.0: How it is defined from a sociotechnical perspective and how much sustainability it includes–A literature review. *Journal of Cleaner Production, 259*, p. 120856.
17. Gorecky, D., Schmitt, M., Loskyll, M. and Zühlke, D., 2014, July. Human-machine interaction in the industry 4.0 era. In *2014 12th IEEE International Conference on Industrial Informatics (INDIN), Porto Alegre, Brazil, (27-30 July), ISBN: 9781479949052* (pp. 289–294), DOI: 10.1109/INDIN.2014.6945523
18. Birkel, H.S., Veile, J.W., Müller, J.M., Hartmann, E. and Voigt, K.I., 2019. Development of a risk framework for industry 4.0 in the context of sustainability for established manufacturers. *Sustainability, 11(2)*, p. 384.
19. Waschull, S., Bokhorst, J.A., Molleman, E. and Wortmann, J.C., 2020. Work design in future industrial production: Transforming towards cyber-physical systems. *Computers, 139*, p. 105679.
20. Wang, B., Hu, S.J., Sun, L. and Freiheit, T., 2020. Intelligent welding system technologies: State-of-the-art review and perspectives. *Journal of Manufacturing Systems, 56*, pp. 373–391.
21. Papadonikolaki, E., van Oel, C. and Kagioglou, M., 2019. Organising and managing boundaries: A structurational view of collaboration with building information modelling (BIM). *International Journal of Project Management, 37(3)*, pp. 378–394.
22. Montresor, S. and Vezzani, A., 2015. The production function of top R&D investors: Accounting for size and sector heterogeneity with quantile estimations. *Research Policy, 44(2)*, pp. 381–393.
23. Caruso, L., 2018. Digital innovation and the fourth industrial revolution: Epochal social changes? *AI & Society, 33(3)*, pp. 379–392.
24. Bernstein, E.S. and Turban, S., 2018. The impact of the 'open' workspace on human collaboration. *Philosophical Transactions of the Royal Society B: Biological Sciences, 373(1753)*, p. 20170239.

25. Cascio, W.F. and Montealegre, R., 2016. How technology is changing work and organizations. *Annual Review of Organizational Psychology and Organizational Behavior, 3*, pp. 349–375.
26. Bucci, S., Schwannauer, M. and Berry, N., 2019. The digital revolution and its impact on mental health care. *Psychology and Psychotherapy: Theory, Research and Practice, 92*(2), pp. 277–297.
27. Bhui, K., Dinos, S., Galant-Miecznikowska, M., de Jongh, B. and Stansfeld, S., 2016. Perceptions of work stress causes and effective interventions in employees working in public, private and non-governmental organisations: A qualitative study. *BJPsych Bulletin, 40*(6), pp. 318–325.
28. Sirgy, M.J. and Lee, D.J., 2018. Work-life balance: An integrative review. *Applied Research in Quality of Life, 13*(1), pp. 229–254.
29. Gerhardt, C., Stocker, D., Looser, D., grosse Holtforth, M. and Elfering, A., 2019. Well-being and health-related interventions in small-and medium-sized enterprises: A meta-analytic review. *Zeitschrift für Arbeitswissenschaft, 73*(3), pp. 285–294.
30. Pfeiffer, S., 2017. The vision of "Industrie 4.0" in the making—A case of future told, tamed, and traded. *Nanoethics, 11*(1), pp. 107–121.
31. Bakotić, D., 2016. Relationship between job satisfaction and organisational performance. *Economic Research-Ekonomska Istraživanja, 29*(1), pp. 118–130.
32. Krzywdzinski, M., 2017. Automation, skill requirements and labour-use strategies: High-wage and low-wage approaches to high-tech manufacturing in the automotive industry. *New Technology, Work and Employment, 32*(3), pp. 247–267.
33. Dalhammar, C., 2016. Industry attitudes towards ecodesign standards for improved resource efficiency. *Journal of Cleaner Production, 123*, pp. 155–166.
34. Milios, L., 2018. Advancing to a circular economy: Three essential ingredients for a comprehensive policy mix. *Sustainability Science, 13*(3), pp. 861–878.
35. Marchi, B. and Zanoni, S., 2017. Supply chain management for improved energy efficiency: Review and opportunities. *Energies, 10*(10), p. 1618.
36. Javied, T., Rackow, T. and Franke, J., 2015. Implementing energy management system to increase energy efficiency in manufacturing companies. *Procedia CIRP, 26*, pp. 156–161.
37. Korhonen, J., Honkasalo, A. and Seppälä, J., 2018. Circular economy: The concept and its limitations. *Ecological Economics, 143*, pp. 37–46.
38. Sodiq, A., Baloch, A.A., Khan, S.A., Sezer, N., Mahmoud, S., Jama, M. and Abdelaal, A., 2019. Towards modern sustainable cities: Review of sustainability principles and trends. *Journal of Cleaner Production, 227*, pp. 972–1001.
39. Vidhi, R. and Shrivastava, P., 2018. A review of electric vehicle lifecycle emissions and policy recommendations to increase EV penetration in India. *Energies, 11*(3), p. 483.
40. Ardente, F., Peiró, L.T., Mathieux, F. and Polverini, D., 2018. Accounting for the environmental benefits of remanufactured products: Method and application. *Journal of Cleaner Production, 198*, pp. 1545–1558.

41. Attaran, M., 2017. The rise of 3-D printing: The advantages of additive manufacturing over traditional manufacturing. *Business Horizons*, 60(5), pp. 677–688.
42. Ligon, S.C., Liska, R., Stampfl, J., Gurr, M. and Mülhaupt, R., 2017. Polymers for 3D printing and customized additive manufacturing. *Chemical Reviews*, 117(15), pp. 10212–10290.
43. Holmström, J., Liotta, G. and Chaudhuri, A., 2017. Sustainability outcomes through direct digital manufacturing-based operational practices: A design theory approach. *Journal of Cleaner Production*, 167, pp. 951–961.
44. Franco, D., Ganga, G.M.D., de Santa-Eulalia, L.A. and Godinho Filho, M., 2020. Consolidated and inconclusive effects of additive manufacturing adoption: A systematic literature review. *Computers & Industrial Engineering*, 148, p. 106713.
45. Todeschini, B.V., Cortimiglia, M.N., Callegaro-de-Menezes, D. and Ghezzi, A., 2017. Innovative and sustainable business models in the fashion industry: Entrepreneurial drivers, opportunities, and challenges. *Business Horizons*, 60(6), pp. 759–770.
46. Davis, J., Malkani, H., Dyck, J., Korambath, P. and Wise, J., 2020. Cyberinfrastructure for the democratization of smart manufacturing. In *Smart Manufacturing, ISBN: 9780128200278, Elsevier* (pp. 83–116). DOI: https://doi.org/10.1016/C2018-0-05201-9.
47. Gomber, P., Koch, J.A. and Siering, M., 2017. Digital finance and FinTech: Current research and future research directions. *Journal of Business Economics*, 87(5), pp. 537–580.
48. Gabriel, M. and Pessl, E., 2016. Industry 4.0 and sustainability impacts: Critical discussion of sustainability aspects with a special focus on future of work and ecological consequences. *Annals of the Faculty of Engineering Hunedoara*, 14(2), p. 131.
49. Besutti, R., de Campos Machado, V. and Cecconello, I., 2019. Development of an open source-based manufacturing execution system (MES): Industry 4.0 enabling technology for small and medium-sized enterprises. *Scientia Cum Industria*, 7(2), pp. 1–11.
50. Tsai, W.H., Lan, S.H. and Huang, C.T., 2019. Activity-based standard costing product-mix decision in the future digital era: Green recycling steel-scrap material for steel industry. *Sustainability*, 11(3), p. 899.
51. Greenhalgh, T., Jackson, C., Shaw, S. and Janamian, T., 2016. Achieving research impact through co-creation in community-based health services: Literature review and case study. *The Milbank Quarterly*, 94(2), pp. 392–429.
52. Raj, A., Dwivedi, G., Sharma, A., de Sousa Jabbour, A.B.L. and Rajak, S., 2020. Barriers to the adoption of industry 4.0 technologies in the manufacturing sector: An inter-country comparative perspective. *International Journal of Production Economics*, 224, p. 107546.
53. Ghafoorpoor Yazdi, P., Azizi, A. and Hashemipour, M., 2018. An empirical investigation of the relationship between overall equipment efficiency (OEE) and manufacturing sustainability in industry 4.0 with time study approach. *Sustainability*, 10(9), p. 3031.

54. Li, Y. and Wang, X., 2018. Risk assessment for public–private partnership projects: Using a fuzzy analytic hierarchical process method and expert opinion in China. *Journal of Risk Research*, 21(8), pp. 952–973.
55. Crouch, C., 2019. Inequality in post-industrial societies. *Structural Change and Economic Dynamics*, 51, pp. 11–23.
56. Xin-Gang, Z., Gui-Wu, J., Ang, L. and Yun, L., 2016. Technology, cost, a performance of waste-to-energy incineration industry in China. *Renewable and Sustainable Energy Reviews*, 55, pp. 115–130.
57. Subramaniam, S.K., Husin, S.H., Singh, R.S.S. and Hamidon, A.H., 2009. Production monitoring system for monitoring the industrial shop floor performance. *International Journal of Systems Applications, Engineering & Development*, 3(1), pp. 28–35.
58. Haque, B. and James-Moore, M., 2004. Applying lean thinking to new product introduction. *Journal of Engineering Design*, 15(1), pp. 1–31.
59. Mrugalska, B. and Wyrwicka, M.K., 2017. Towards lean production in industry 4.0. *Procedia Engineering*, 182, pp. 466–473.
60. Arruñada, B. and Vázquez, X.H., 2006. When your contract manufacturer becomes your competitor. *Harvard Business Review*, 84(9), p. 135.
61. Handoo, S., Arora, V., Khera, D., Nandi, P.K. and Sahu, S.K., 2012. A comprehensive study on regulatory requirements for development and filing of generic drugs globally. *International Journal of Pharmaceutical Investigation*, 2(3), p. 99.
62. Haller, S., Karnouskos, S. and Schroth, C., 2008, September. The internet of things in an enterprise context. . In *Future Internet–FIS 2008: First Future Internet Symposium, FIS 2008 Vienna, Austria, September 29–30, 2008 Revised Selected Papers 1*, pp. 14–28. Springer Berlin Heidelberg.
63. Hofmann, E. and Rüsch, M., 2017. Industry 4.0 and the current status as well as future prospects on logistics. *Computers in Industry*, 89, pp. 23–34.
64. Butt, J., 2020. Exploring the interrelationship between additive manufacturing and industry 4.0. *Designs*, 4(2), p. 13.
65. Russo, M., 1985. Technical change and the industrial district: The role of interfirm relations in the growth and transformation of ceramic tile production in Italy. *Research Policy*, 14(6), pp. 329–343.
66. Zhong, R.Y., Xu, X., Klotz, E. and Newman, S.T., 2017. Intelligent manufacturing in the context of industry 4.0: A review. *Engineering*, 3(5), pp. 616–630.
67. Büchi, G., Cugno, M. and Castagnoli, R., 2020. Smart factory performance and industry 4.0. *Technological Forecasting and Social Change*, 150, p. 119790.
68. Xiang, Z., Schwartz, Z., Gerdes Jr, J.H. and Uysal, M., 2015. What can big data and text analytics tell us about hotel guest experience and satisfaction? *International Journal of Hospitality Management*, 44, pp. 120–130.
69. Balletti, C., Ballarin, M. and Guerra, F., 2017. 3D printing: State of the art and future perspectives. *Journal of Cultural Heritage*, 26, pp. 172–182.
70. Roy, M. and Roy, A., 2019. Nexus of internet of things (IoT) and big data: Roadmap for smart management systems (SMgS). *IEEE Engineering Management Review*, 47(2), pp. 53–65.

71. Pishva, D., 2017, February. Internet of things: Security and privacy issues and possible solution. In *2017 19th International Conference on Advanced Communication Technology (ICACT), Pyeong Chang, Korea (South), INSPEC Accession Number: 16777286, IEEE* (pp. 797–808). **DOI:** 10.23919/ICACT.2017.7890229

72. Zhao, D. and Rosson, M.B., 2009, May. How and why people Twitter: The role that micro-blogging plays in informal communication at work. In *Proceedings of the ACM 2009 International Conference on Supporting Group Work, Association for Computing Machinery, New York, United States* (pp. 243–252). https://doi.org/10.1145/1531674.1531710.

73. LeCun, Y., Bengio, Y. and Hinton, G., 2015. Deep learning. *Nature, 521*(7553), pp. 436–444.

74. Rehman, U. and Cao, S., 2015, October. Augmented reality-based indoor navigation using google glass as a wearable head-mounted display. In *2015 IEEE International Conference on Systems, Man, and Cybernetics, Hongkong China, INSPEC Acession Number: 15718753* (pp. 1452–1457) DOI: 10.1109/SMC.2015.257.

75. Darvishi, H., Ciuonzo, D., Eide, E.R. and Rossi, P.S., 2020. Sensor-fault detection, isolation and accommodation for digital twins via modular data-driven architecture. *IEEE Sensors Journal, 21*(4), pp. 4827–4838.

76. Segel, E. and Heer, J., 2010. Narrative visualization: Telling stories with data. *IEEE Transactions on Visualization and Computer Graphics, 16*(6), pp. 1139–1148.

77. Aceto, G., Persico, V. and Pescapé, A., 2020. Industry 4.0 and health: Internet of things, big data, and cloud computing for healthcare 4.0. *Journal of Industrial Information Integration, 18*, p. 100129.

78. Minoli, D. and Occhiogrosso, B., 2018. Blockchain mechanisms for IoT security. *Internet of Things, 1*, pp. 1–13.

79. Grubesic, T.H. and Murray, A.T., 2004. Waiting for broadband: Local competition and the spatial distribution of advanced telecommunication services in the United States. *Growth and Change, 35*(2), pp. 139–165.

80. Fernando, N., Loke, S.W. and Rahayu, W., 2013. Mobile cloud computing: A survey. *Future Generation Computer Systems, 29*(1), pp. 84–106.

81. Endsley, M.R. and Kaber, D.B., 1999. Level of automation effects on performance, situation awareness and workload in a dynamic control task. *Ergonomics, 42*(3), pp. 462–492.

82. Coroama, V.C., Moberg, Å. and Hilty, L.M., 2015. Dematerialization through electronic media? In *ICT Innovations for Sustainability*, Springer, ISBN: 9783319092270 (pp. 405–421). https://doi.org/10.1007/978-3-319-09228-7_24.

83. Nahavandi, S., 2019. Industry 5.0—A human-centric solution. *Sustainability, 11*(16), p. 4371.

84. Rajkumar, R., Lee, I., Sha, L. and Stankovic, J., 2010, June. (DAC' 10 Proceedings of the 47th Design automation conference, ISBN: 9781450300025, (pp. 731-736). DOI: http: //doi.org/10.1145/1837274.

85. Huang, M.H., Rust, R. and Maksimovic, V., 2019. The feeling economy: Managing in the next generation of artificial intelligence (AI). *California Management Review, 61*(4), pp. 43–65.

86. Hacioglu, U. and Sevgilioglu, G., 2019. The evolving role of automated systems and its cyber-security issue for global business operations in Industry 4.0. *International Journal of Business Ecosystem & Strategy (2687–2293)*, 1(1), pp. 1–11.

87. Rao, S.K. and Prasad, R., 2018. Impact of 5G technologies on industry 4.0. *Wireless Personal Communications*, 100(1), pp. 145–159.

88. Prasetio, A.P., Luturlean, B.S. and Agathanisa, C., 2019. Examining employee's compensation satisfaction and work stress in a retail company and its effect to increase employee job satisfaction. *International Journal of Human Resource Studies*, 9(2), pp. 239–265.

89. Lee, S.M., Lee, D. and Kim, Y.S., 2019. The quality management ecosystem for predictive maintenance in the industry 4.0 era. *International Journal of Quality Innovation*, 5(1), pp. 1–11.

90. Schimek, R.S., 2016. IoT case studies: Companies leading the connected economy. In *American International Group* (pp. 1–16).

91. Fettermann, D.C., Cavalcante, C.G.S., Almeida, T.D.D. and Tortorella, G.L., 2018. How does industry 4.0 contribute to operations management? *Journal of Industrial and Production Engineering*, 35(4), pp. 255–268.

92. Klingenberg, C. and Antunes, J., 2017. Industry 4.0: What makes it a revolution. *EurOMA 2017*, pp. 1–11.

93. Hannibal, M. and Knight, G., 2018. Additive manufacturing and the global factory: Disruptive technologies and the location of international business. *International Business Review*, 27(6), pp. 1116–1127.

94. Unger, H., Börner, F. and Müller, E., 2017. Context related information provision in industry 4.0 environments. *Procedia Manufacturing*, 11, pp. 796–805.

95. Motyl, B., Baronio, G., Uberti, S., Speranza, D. and Filippi, S., 2017. How will change the future engineers' skills in the Industry 4.0 framework? A questionnaire survey. *Procedia Manufacturing*, 11, pp. 1501–1509.

96. Geissdoerfer, M., Morioka, S.N., de Carvalho, M.M. and Evans, S., 2018. Business models and supply chains for the circular economy. *Journal of Cleaner Production*, 190, pp. 712–721.

97. Küsters, D., Praß, N. and Gloy, Y.S., 2017. Textile learning factory 4.0–preparing Germany's textile industry for the digital future. *Procedia Manufacturing*, 9, pp. 214–221.

98. Iyer, A., 2018. Moving from industry 2.0 to industry 4.0: A case study from India on leapfrogging in smart manufacturing. *Procedia Manufacturing*, 21, pp. 663–670.

99. Nagy, J., Oláh, J., Erdei, E., Máté, D. and Popp, J., 2018. The role and impact of industry 4.0 and the internet of things on the business strategy of the value chain—The case of Hungary. *Sustainability*, 10(10), p. 3491.

100. nd . Turber, S. and Smiela, C., 2014, A Business Model Type for the Internet of Things. Proceedings of the Twenty Second European Conference on Information System (ECIS), Tel, Aviv, Israel (June 9-11), ISBN: 9780991556700 (pp. 1–10). DOI: http://aisel.aisnet.org/ecis2014/proceedings/track05/4.

101. Kaliczyńska, M. and Dąbek, P., 2015. Value of the internet of things for the industry–an overview. *Mechatronics-Ideas for Industrial Application*, Springer, ISBN: 9783319109909, pp. 51–63. DOI : https://doi.org/10.1007/978-3-319-10990-9_6.

102. Gajdzik, B., Grabowska, S. and Saniuk, S., 2021. A theoretical framework for industry 4.0 and its implementation with selected practical schedules. *Energies*, *14*(4), p. 940.
103. Kumar, A., 2004. Mass customization: Metrics and modularity. *International Journal of Flexible Manufacturing Systems*, *16*(4), pp. 287–311.

Chapter 2

Smart sensors and actuators in Industry 4.0

Hema Gurung Ravi Kant

2.1 INTRODUCTION

The fourth industrial revolution is Industry 4.0, also called smart manufacturing and Internet of Things (IoT). It describes the continuous improvement towards automation, interconnectivity, machine learning, and better control over the production process. This makes mass production easier with an improved and cost-effective end product. In simple words, it's a synergetic combination of physical product and smart digital technology which helps in achieving a better interconnected ecosystem. The one difference between the traditional approach and Industry 4.0 is the service. Traditional manufacturing lacks automated control and monitoring capabilities [1], hence applications are isolated and stand-alone [2]. It includes a number of different and independent steps, starting from product development, manufacturing, marketing, and finally supplying the product to the customer [3]. The main discrepancy of traditional manufacturing is that it does not include any integration of physical and digital systems [4]. Traditional manufacturing does not include advanced strategies such as agile, intelligent, and flexible manufacturing [5–7], which makes it lag behind the modern manufacturing sector. Whereas Industry 4.0 involves all the advanced manufacturing strategies resulting in the interaction between machine and product possible without or with minimal human intervention [2, 4]. The Industry 4.0 based manufacturing process focuses mainly on product development, sustainability, lean manufacturing, and enhancing strategic management. In the development of modern society, the manufacturing industry plays an important role and Industry 4.0 makes the industries smarter and more intelligent. The smart factories/industries are using various advanced technologies like cloud and cognitive computing, 3D/4D printing, advanced robotics, and many more. The use of advanced and intelligent industrial robots together with computer numerical control machines aided flexibility in the manufacturing process [8–10]. Similarly, computer integrated manufacturing became possible [11, 12] only with the help of computer aided design and computer aided processing planning. Further, the IoT enabled manufacturing to go for digital transformation

DOI: 10.1201/9781003246466-2

of various scenarios such as automation, efficient production, competitive advantages, and customer focus [13, 14]. Researchers have also investigated the contribution of IoT in Industry 4.0 related to its connectivity, flexibility, information sharing, and transparency. In the manufacturing industries, the efficient use of IoT is conceivable by using smart sensors and actuators, making manufacturing industries efficient for Industry 4.0. The smart sensors yield better performance, higher integration, and also have the ability to sense multiple parameters. They also provide intelligent as well as secure and safe networking [15]. Similarly, in industries, 4.0 smart actuators replace manual/human intervention. The beauty of a smart actuator is that it can monitor its own health and check whether the system is receiving adequate input signal for the job or needs further modification.

This chapter presents the different aspects related to Industry 4.0, which are as follows:

- Relevance and importance of the advanced technologies pertaining to Industry 4.0.
- Identifying the variation between traditional and smart factories.
- Different actuators and sensor technology available for Industry 4.0
- The digital transformation of information using the fusion of smart sensors, actuators, and IoT.

2.2 INDUSTRY 4.0: RELEVANCE AND IMPLICATIONS

Nowadays, many companies and industries are using Industry 4.0 to satisfy customer demands. Researchers are continuously focusing on modern and smart technology to make the manufacturing process more advanced. Industry 4.0 is an extension of the currently existing manufacturing process. The first industrial revolution, i.e., Industry 1.0 was a manual labour based industry where steam engines and water power were used to increase productivity. This is also called the mechanization of industries.

Industry 2.0, the second industrial revolution, also called the technological revolution, was a period of economic growth, with the utilization of electricity and assembly line production. In this revolution, productivity increased but at the same time brought unemployment as human efforts were replaced by machines.

The third industrial revolution, i.e., Industry 3.0, the digital revolution, boosted the production process and partially automated the product by introducing digital computer and memory programmable controls. In this phase, more importance was given to mass production. The use of digital logic, MOS transistors, integrated circuit chips, and their deriving technologies, which include the Internet, microprocessors, and digital cellular phones, have also increased.

Next comes the fourth industrial revolution, i.e., Industry 4.0, which promotes the computerization of manufacturing processes. Even though the fourth revolution is now coming into existence, it is important to note that some of the factories are still only at the beginning of the Industry 3.0 phase. Industry 4.0 basically represents the blending of physical quantities with advanced digital technologies, like artificial intelligence, 3D/4D printing, advanced robotic systems, cloud computing, and many more. Since both systems and technologies are interconnected, they can communicate with each other and hence can act accordingly to give a better product – with less human effort. In short, Industry 4.0 makes the manufacturing process more intelligent, flexible, cost effective, adaptable, and reactive.

Industry 4.0 is known for its four design principles:

Connection and communication – Machines, sensors, devices, and people are interconnected and hence can easily communicate with each other using IoT.

Information transparency – The information provided by Industry 4.0 is clear and transparent. Hence, an operator can easily take decisions.

Technical guidance – Technical guidance of the systems assists human beings in making decisions, solving problems, and also guides the human during difficult and unsafe tasks.

Distributed decisions – Physical cyber systems enable the systems to take their own decision and can also perform the task independently, if required.

2.3 SMART FACTORIES

The smart factory is defined as the factory where various Industry 4.0 technologies like digital technology, smart computing, IoT, etc. are combined with the production process and its operation. It is a new phase of the industrial revolution that creates a more opportunistic system for the manufacturing industry.

A smart factory is characterized with the help of:

Smart sensors – These have self-learning and decision-making capability. They can sense the physical parameters or collect information from the environment and can make decisions depending on the available information.

Smart actuators – The actuator is manipulated or controlled by computers or programmable logic; hence, eliminating the need for a human operator.

Internet of Things (IoT) – This allows the interconnection between machines, devices, sensors, actuators, and processors that are connected by communication systems. It can capture the data and also remotely monitor and control the production. It also helps in exchanging the data between customers and machines.

Cloud computing – This makes data sharing, storing, and processing very flexible. Interconnected devices and machines can upload large amounts of data. The stored data is then refined for real time applications and also used to provide feedback.

Robotics – This helps in performing repetitive, dangerous and dirty tasks. It can also perform tasks that require a very high level of accuracy. A smart factory requires an autonomous robot or cognitive robot or both to make the factory smarter and more intelligent.

All these technologies and many more are employed on the production and assembly lines of the manufacturing industries [16].

2.3.1   Benefits of smart factory

The smart factory is a flexible manufacturing system that can self-analyse the performance of the system and, if required, subsequently correct the process. It can also easily adapt to or learn from new environments and run the entire production processes autonomously and smoothly.

The main benefits of the smart factories are as follows:

(1) **Asset performance** – Smart factories generate large quantities of data. This data is continuously analysed to verify the asset performance by the processor. If some discrepancies occur, then it can undergo some kind of self-corrective optimization. In fact, the self-correction capability of the smart factory makes it superior to the traditional factory.

(2) **Quality** – The self-optimization characteristics of the smart factory help in detecting and predicting the quality defect. They also help in identifying the reasons behind the poor quality of the product and trying to remove/rectify the defects. A more optimized quality process could lead to a better quality product with fewer defects.

(3) **Cost effective** – The optimized process of production is a very cost-effective process. It yields a better end product as it includes a better quality of production process that integrates the data received from physical systems, various operational devices, and human resources. All this information within the manufacturing industry is used to drive the manufacturing process, its maintenance, tracking of record, digitization of operations, and many more. These make the manufacturing industry more efficient with reduced production time and improved quality products. These advantages contribute to reducing the overall cost and leading the smart factory to a better position in this competitive world.

(4) **Safety and sustainability** – Smart factories also provide labour safety and environmental sustainability. The self-sensing and self-correction

capability of the smart factory make it more intelligent and autonomous. Hence, it causes less human intervention and less human
error, reducing industrial accidents that cause injury to the humans.
Further, its operational efficiencies provide greater environmental
sustainability.

Hence, we can say a smart factory incorporates all the advanced
technology starting from smart sensors and actuators, intelligent
robots, advanced digital systems, and many more, making the manufacturing industry more efficient and productive.

2.4 SMART FACTORY VS. TRADITIONAL FACTORY

Nowadays, customers are demanding varieties of products and manufacturing industries are trying their best to satisfy the customer demands. Hence,
in order to handle large product varieties, manufacturing industries must
have the capability of customizing the product type as well as product capacity [17]. Further, to satisfy the customer requirements and also to overcome
the different challenges, the manufacturing processes should have enough
functionality and scalability with a good connection between customers and
suppliers. Smart factories have all these qualities and are more intelligent
and smarter than traditional factories. The conventional factories lack self-
sensing and self-decision-making capability and hence are unable to monitor and control the manufacturing process automatically, causing inefficient
production of the modified products [18]. The integration of production
systems, or we can say the integration between real and virtual systems, is
lower in traditional factories. As a result, there is poor reuse of the system.

In the earlier era, products were hand-hade, taking a tremendous amount
of time to complete a single product. The industrial revolutions made the
manufacturing process more advanced – i.e., from man-made (Industry
1.0) to industry man-machine made (Industry 2.0) to a more advanced
one (Industry 3.0) and finally, completely machine made (fully automatic –
Industry 4.0). In an advanced manufacturing process or smart factory, all
the devices, systems, or technology involved are interconnected, allowing
the exchange of information to assess the situation on its own [19, 20]. The
data/information received from the interconnected systems or devices helps
in learning and adapting to the new changes, and therefore can satisfy the
customer needs [21, 22]

The beauty of the smart factory is that the whole manufacturing line can
be customized even at the last minute depending on the product demands,
production specifications, and other settings of the operational devices or
machine [23]. Hence, we can say smart factories make technologies more
accurate and advanced, enhance the product quality, controllability, system
performance, and transparent manufacturing process. The self-adapting
and self-learning features of smart factories make them more friendly and

predictive. Hence, it can circumvent operational downtime and other possible failures in the manufacturing processes.

2.5 SENSORS IN SMART FACTORIES

The key elements of smart factories are intelligent machines, advanced devices, and controllers that can monitor and control the parameters in the manufacturing process. The upgrade from traditional manufacturing to smart factories will encourage the excellent and stable collaboration between machines and systems. In smart factories, future prediction, fault diagnosis, warning generation, status updates, etc. of the machine are possible only from the analysis of the data collected through sensors [24]. Sensors are electrical, electronic, or opto-electrical devices that detect the change in the physical parameters or determine the presence of a particular entity or function, and produce an output signal [25]. The output signal is in some form of energy such as heat, light, motion, or chemical reaction. The data from the environment is collected by the sensor. Then, depending on the requirement, some signal conditioning is done. The corrected data is then implemented into the manufacturing industry to enhance product quality and quantity.

The information received from the sensors is then evaluated and is used for making the decision about the operations that are being carried out [26]. In this era of automation, sensors are used everywhere.

Sensors can be of two types:

- Active and passive sensors
- Analog and digital sensors

This section presents detailed information about the sensors that are commonly used in smart factories.

2.5.1 Active and passive sensors

Active sensor – Active sensors require an external excitation or power signal to do the work. Figure 2.1 (a) depicts a simple example of an active sensor, which is composed of both a transmitter and a receiver. The transmitter emits the electromagnetic energy emitted from the sensors towards the object (house). The transmitted signal interacts with the object and is reflected back to the sensor's receiver.

The following are the properties of an active sensor:

- Contains both transmitter and receiver.
- Suitable for microwave regions of EMR spectrum, where signals can penetrate clouds and are not affected by rain.
- Is independent of solar radiation and can work day and night in all weather.

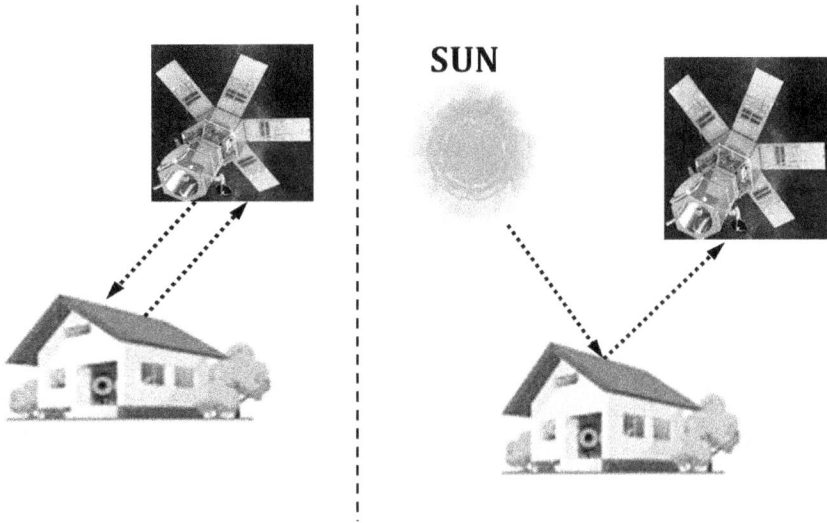

Figure 2.1 Working of active and passive sensor (a) active sensor (b) passive sensor

Passive Sensor – Passive sensors don't require an external excitation or power signal to perform the work. It depends on the external light sources to determine the various characteristics of objects. Figure 2.1 (b) represents a simple example of a passive sensor. The solar energy reflected from the object (house) is received by the sensor. Similarly, the infrared devices are also passive sensors as the excitation signal is generated from infrared radiation that is linked with body temperature. Basically, passive sensors can only be used to detect an object when the naturally occurring energy is available.

The following are the properties of the passive sensor

- It depends on good weather conditions.
- It is suitable for wide band systems.
- It doesn't require high power.

2.5.2 Analog sensor

Analog sensors produce a continuous signal proportional to the measured or input sensed parameters. The continuous output signal produced by the analog sensors is proportional to the measurement. Common examples of analog sensors are light dependent resistor (LDR), pressure sensor, temperature sensor, accelerometer, and many more. LDR absorbs light energy and converts it into a corresponding analog value which lies between 0 and 255. Similarly, the pressure sensors, which are used to measure the amount of pressure applied on the sensor, convert the measured pressure into an analog output signal.

2.5.3 Digital sensor

In digital sensors, data is converted and transmitted digitally and produces a discrete value (0 and 1s). The digital sensor mainly consists of sensors, cables, and transmitters in which the signal measured from the sensor is directly converted into digital form, inside the sensor itself. This is because the digital sensor has an inbuilt signal processing unit like ADC (analog to digital converter), which converts the sensed input signal (analog form) into digital signal. One main advantage of digital sensors is that the digital output of the sensor can be directly interfaced with the digital controller. Further, it can overcome the drawbacks of analog sensors. There is a wide range of different digital sensors like the digital temperature sensor, digital accelerometer, digital pressure sensor, and so on.

Researchers are developing various types of sensors to make the manufacturing process more advanced and bring more automation into smart factories. Smart factories are also using different types of sensors, from basic sensors like temperature, pressure, humidity sensors, etc, to more advanced sensors such as position and product quality sensing [27, 28]. In smart factories, sensors make the manufacturing process more efficient by monitoring or guiding the factory operations as they help in sensing the physical condition, monitoring, and controlling the moving product, controlling the robots, etc. In any manufacturing process, sensors are used mainly to measure pressure, temperature, force, flow, and position. In this section we will discuss the sensors in detail.

2.5.4 Pressure sensor

Pressure sensors are electro-mechanical devices that are used to measure the pressure or force in gases or liquids and provide control signals to the display devices. Basically, pressure sensors detect the pressure change and corresponding to that generates an equivalent electrical signal. Pressure sensors can also measure the change in the atmospheric pressure, as barometric pressure sensors detect the changes in the atmosphere's pressure. This information is used to predict the weather patterns and related changes. Pressure sensors are classified into digital sensors and mechanical sensors as shown in Figure 2.2. The digital pressure sensors are further classified as piezoelectric sensors and strain gauge sensors. The commonly used sensor in the mechanical pressure sensors category is spring scale.

2.5.4.1 Piezoelectric sensor

A piezoelectric sensor uses the piezoelectric effect to measure the changes in pressure, acceleration, temperature, strain, or force and generates an equivalent electrical charge. The main beauty of the piezoelectric sensors is that they don't need any external voltage or current source to operate,

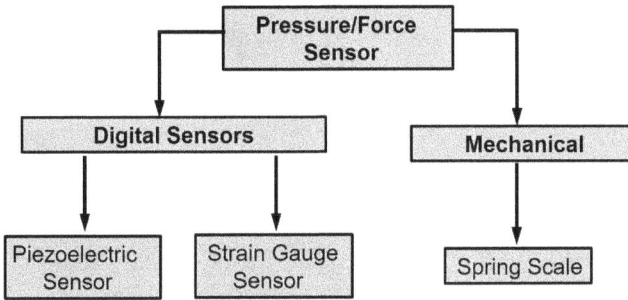

Figure 2.2 Block diagram of the pressure sensor

making them a popular choice for many industrial applications. The word "Piezo" comes from Greek which means "press," so a piezoelectric sensor is used to measure the force produced by compression, using piezoelectric effect. Piezoelectric sensors are highly sensitive and are available in small sizes making them suitable for everyday applications.

2.5.4.2 Working principle

When a pressure or force is applied to the piezoelectric sensor, an equivalent electric charge will generate across the faces of the sensor. The generated charge can be measured as a voltage signal using voltage measuring devices. This measured voltage is proportional to the applied pressure. Further, there also occurs an inverse piezoelectric effect where the shape of the material changes depending on the voltage applied across the piezoelectric material. This happens because applied force causes generation of charges across the materials.

2.5.4.3 Applications of piezoelectric sensor

- Piezoelectric sensors are used in shock detection.
- It is used in flow sensors.
- It is used in accelerometers, microphones, etc.
- It is also used in ultrasound imaging.

2.5.5 Strain gauge sensor

A strain gauge is a sensor that is commonly used in industry to measure the external forces applied on an object.

Strain gauges generate an equivalent electrical signal corresponding to the applied force, pressure torque, etc. The applied force that causes strain is measured with a strain gauge using changes in electrical resistance. The electrical resistance variation which is proportional to the strain experienced by the object can be calculated using the following relation

$$K = \frac{\dfrac{\Delta R}{R}}{\in} \tag{2.1}$$

Where, " ΔR " represents the change in resistance in ohms, "R" is the known resistance, "K" denotes the gauge factor which is usually less than 2 for a metallic foil and " $\in$ " is the strain; need to calculate.

A Wheatstone bridge is commonly used to measure the electrical signal, i.e., change in electrical resistance across the metal foil as shown in Figure 2.3. The Wheatstone bridge generates a voltage signal corresponding to the change in electrical resistance of the strain gauge.

The output voltage of the Wheatstone bridge can be given by the formula:

$$V_o = V_{in} \left(\frac{R_2}{R_G + R_2} - \frac{R_1}{R_1 + R_3} \right) \tag{2.2}$$

Where, V_o is the measured output voltage and V_{in} is the bridge excitation voltage.

The strain measured using strain gauge is then used to determine the stress experienced by an object using Hooke's law as

$$\sigma = E \in \tag{2.3}$$

Where, "σ" represents the applied stress and "E" is the material's modulus of elasticity.

2.5.5.1 Working principle

When a force is applied to an external object, it causes a change in strain. The change in strain which gives information about the amount of force applied to an object is measured using strain gauge as illustrated in Eqn. (2.3). When any force "F" is applied to a beam, it deforms and hence causes changes in the strain. The strain gauge fixed on the top of the beam is used to measure this strain. Therefore, in order to measure the strain using strain gauge, it must be connected to an electrical circuit that is capable of measuring the smallest changes in resistance associated with strain. A Wheatstone bridge is used to measure the change in the electrical resistance of the device. In a Wheatstone bridge configuration, an input excitation voltage is applied across the circuit, and the output voltage is measured across two points in the middle of the bridge as illustrated in Figure 2.3.

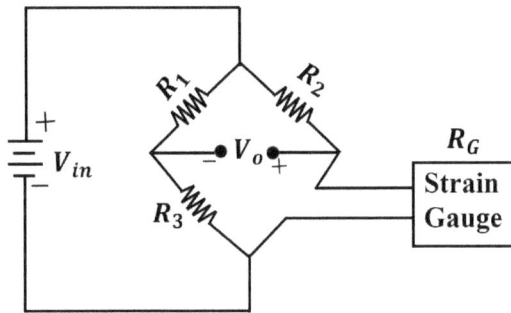

Figure 2.3 Force calculation using strain gauge with the help of Wheatstone bridge

When no force is acting on the gauge, the Wheatstone Bridge is balanced and the output voltage will be zero.

Now, when the load is applied to the material, the strain gauge also deforms with the material. Hence, it causes a change in strain which also results in a change in the resistance of the strain gauge. This makes the bridge unbalance and hence, causes a change in the output voltage. From the measured voltage, the strain of the gauge can be calculated, which is further used to measure the applied forces.

2.5.5.2 Characteristics of strain gauge

There are a few characteristics of strain gauge that are needed for proper functioning of the device.

2.5.5.2.1 Sensitivity

Sensitivity is defined as the smallest absolute amount of change in the value of the strain that can be detected by the sensor. The choice of a strain gauge for a particular application highly depends upon the degree of sensitivity of the sensor. Therefore, to measure the strain, one has to accurately measure the very small changes in resistance. It is observed that the selection of a gauge with a very high sensitivity increases the complexity of the measuring method.

2.5.5.2.2 Accuracy

Accuracy is defined as the amount of uncertainty in a measurement with respect to an absolute standard. Basically, it measures the degree of closeness w.r.t. true value. In mechanical strain gauge lost motion such as temperature changes, backlash gear train, slippage, deflection of components, etc., results in inaccuracies.

2.5.5.2.3 Range

The range of a strain gauge is defined as the maximum strain that can be measured without resetting or replacing the strain gauge. It is important to note that the range and the sensitivity of the device are interrelated. Hence, strain gauge with very high sensitivity responses to a small strain.

2.5.5.2.4 Applications

Strain gauges are commonly used in

- Aerospace.
- Vibration and torque measurement.
- Bending and deflection measurement.
- Detect failures in structures like bridges, buildings, and much more.

2.5.6 Spring scale

It is a type of weighing scale that gives the relations between applied load and spring deformation. The relation between force applied and spring deflection is usually linear and is based on Hooke's law. One end of the helical spring is attached to the casing and is fixed. Another end is attached to a crossbar to which a hook is attached for carrying the load. The pinion to which the indicating pointer is attached is pivoted on the casing and meshes with the rack. This is pivotally connected to the crossbar and is pressed into contact with the pinion by the rack spring.

2.5.6.1 Working principle

When the load is applied, the spring stretches depending on the load applied. As the spring deformed, it causes the movement of the crossbar and the attached rack. As a result, the pinion rotates and, accordingly, the load indicating the pointer rotates. The marking of the dial is equally spaced and its scale units depend on the spring stiffness; the spring having higher stiffness has large scale units and a good load carrying capacity.

2.5.6.2 Applications

- Spring balances are widely used commercially.
- Used to carry heavy loads such as trucks, storage silos, etc.

2.5.7 Temperature sensor

A temperature sensor is a device which is used to measure the temperature of air, liquid, and solid matter. It measures the degree of hotness or coldness

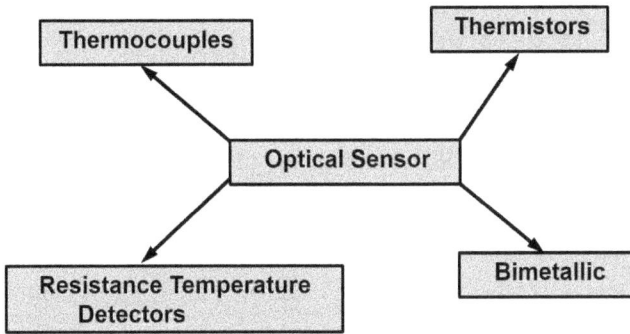

Figure 2.4 Block diagram representation of the temperature sensor

of an object and converts the measured temperature into an electrical signal. There are different types of temperature sensors available in the market as shown in Figure 2.4. Each temperature sensor uses different principles and technologies to measure the temperature.

2.5.7.1 Different types of temperature sensor

2.5.7.1.1 Thermistor

The term thermistor comes from two words, i.e., "thermal" and "resistor". It is a thermally sensitive resistor whose physical appearance changes with the change in temperature.

Thermistors are made up of ceramic materials like oxides of nickel, manganese, and coated cobalt in a glass, making them easily deformable. The temperature measured by the thermistor is linearly related with the resistance of the thermistor and can be given as:

$$\Delta R = K \Delta T \tag{2.4}$$

Here, ΔR represents the change in the resistance value, K is the temperature coefficient of first order and ΔT denotes the change in temperature. Depending on the value of the temperature coefficient thermistor can be either referred as "negative temperature coefficient" (NTC) or "positive temperature coefficient" (PTC)

Positive temperature coefficient (PTC) – In PTC the coefficient of temperature is positive and hence, with the increase in temperature, the value of resistance also increases. The PTC type thermistor is further classified into two types. The first one is (a) silistors – made up of silicon and have linear relationship between the thermistor's temperature and resistance, and the second type is (b) the switching type of PTC – here, initially the thermistor behaves like NTC, where the thermistor's resistance decreases with the increase in temperature. But once the temperature reaches the transition point of the device known as Curie temperature, the resistance of the thermistor again increases with the increase in temperature.

Negative temperature coefficient (NTC) – Compared to PTC, NTC has a negative temperature coefficient and hence, the resistance of the thermistor decreases with the increase in temperature. The NTC type of thermistor is commonly used because it can easily be implemented into any device where temperature plays an important role. Depending on the materials used in the construction of NTC and the process of production, NTC is divided into three groups:

- **Bead thermistors** – made by taking the lead wires of platinum alloy which are directly interfaced to the ceramic body. It has faster response time, better stability, and can operate at very high temperatures.
- **Disk and chip thermistors** – manufactured using metal surface contact. As compared to the bead thermistor, disk and chip thermistors are large and hence have a slow response time. However, the current carrying capacity of this thermistor is better than the bead thermistor. This is because the power dissipation of this thermistor is proportional to the square of the flowing current.
- **Glass encapsulated thermistors** – these are more stable and are protected from environmental changes and are used for the temperature above 150°C.

2.5.7.1.1.1 WORKING PRINCIPLE OF THERMISTOR

The working principle of a thermistor is that the resistance of the thermistor changes with the change in the thermistor's temperature. The resistance value can be measured using ohmmeter, which is connected in series with the battery and the meter, as shown in Figure 2.5. The change in the resistance of the thermistor with the temperature change depends on the materials that are used in the construction of the thermistors. Depending on the thermistor's materials, the coefficient of temperature either increases or decreases, depending on the temperature coefficient as given in Eqn. (2.4). The resistance either increases or decreases with the temperature.

2.5.7.1.1.2 APPLICATIONS OF THERMISTORS

- Thermistors are compact and hence can be used in digital thermometers.
- Household electronics: washing machine, electric cooker, etc.
- Air-conditioning: room temperature monitoring, i.e., to increase or decrease the amount of heat required.
- Industrial process control.
- Thermistors are also used for time delay and switching applications.

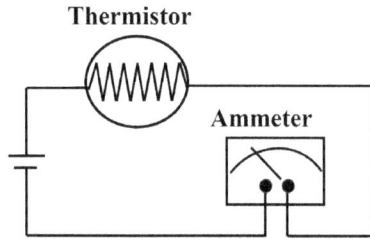

Figure 2.5 Temperature calculation using thermistor with the help of ammeter

2.5.7.1.2 Thermocouples

A thermocouple is a device or sensor used to measure the temperature. It is commonly used in industry, homes, offices, etc. It is a cost-effective temperature sensor which can measure a wide range of temperatures. A thermocouple is also defined as a thermal junction because it is based on the thermoelectric effect where temperature difference is directly converted to electric voltage.

2.5.7.1.2.1 WORKING PRINCIPLE

Thermocouples which are used for temperature measurement consist of two plates of different metals. The welded one end of the plates creates a junction and the other end of the plates is free which is connected to the measuring device, as shown in Figure 2.6. The junction is the point that is placed on the targeted surface whose temperature has to be measured. This junction is called the "hot" junction, whereas, the other end of the plate is kept at low or ambient temperature and is called the "cold" or "reference" junction. When the hot junction experiences a temperature change, i.e., there occurs some temperature difference between two metal plates, a voltage is created between the plates of the thermocouples. The induced voltage, which is measured by the voltmeter, is a function of the temperature between the junctions. Hence, by measuring the voltage, we can calculate the temperature of the system.

2.5.7.1.2.2 TYPES OF THERMOCOUPLES

There are different types of thermocouples available in the market. They are segregated depending on the different combinations of alloy that are used in the construction of thermocouples. Each thermocouple has different characteristics, like different temperature ranges, vibration and chemical resistance, durability, etc. Depending on the application, cost, temperature range, stability, etc., the particular type of thermocouple can be chosen. Some of the commonly used thermocouples are discussed below.

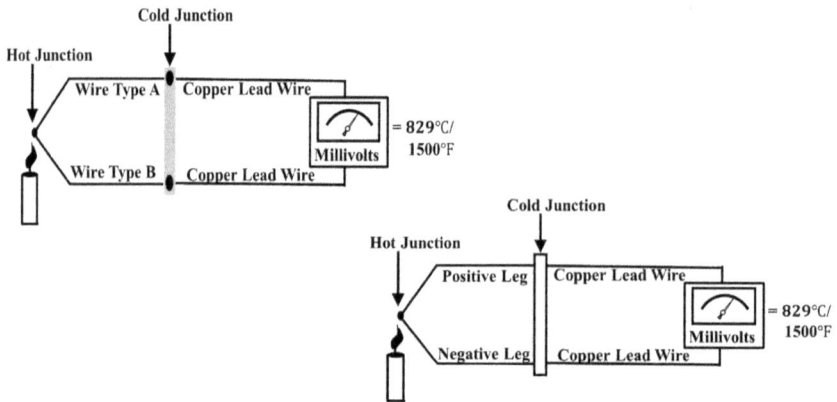

Figure 2.6 Temperature measurement of hot junction in terms of voltage using thermocouple

2.5.7.1.2.2.1 Type K thermocouples K type thermocouple is the most common type of thermocouple. It has the widest temperature measuring range, which can vary from -270°C to 1,260°C or from –454°F to 2,300°F. It has two leads, namely positive and negative lead. The positive lead is composed of approximately 90 percent nickel and ten percent chromium and the negative lead is composed of approximately 95 percent aluminium, two percent manganese and one percent silicon. It's a low-cost thermocouple and can be used in those applications where the requirement of temperature sensitivity is $41\frac{\mu V}{°C}$. Thermocouples with a plug and cable can be identified by their respective colour code.

2.5.7.1.2.2.2 Type T thermocouples The type T thermocouples are commonly used to measure low temperature. The sensitivity of this type of thermocouple is nearly $43\frac{\mu V}{°C}$ and can measure the temperature that ranges from –270°C to 370°C . It also has two leads, namely positive and negative lead. The negative lead is composed of constantan, which contains 45 percent nickel and 55 percent copper, and the positive lead is composed of copper.

2.5.7.1.2.2.3 Type J thermocouples Type J thermocouples are also very commonly used thermocouples. It is a low-cost sensor and can measure the temperature that ranges from 0 to 760°C or we can say from $32°F$ to $1,400°F$. The sensitivity of the type J thermocouple is 50 µV/°C. As compared to the K type thermocouple, the J type thermocouple has a smaller temperature range. Again, the type J thermocouple is also composed of two dissimilar metals, the positive lead is made from iron and is coloured black and the negative lead is made from constantan and its coloured red. The overall jacket of the thermocouple is coloured black. This type of thermocouple is not suitable for lower temperature application. However, it can perform best in an oxidizing environment.

2.5.7.1.2.2.4 Type S thermocouples S type thermocouples are used for high temperature application. They can measure the temperature ranging from 0 to $1,600°C$. Like the above-discussed thermocouples, it also has positive and negative leads. The positive lead is made from 90 percent platinum and 10 percent rhodium and the negative lead is made from platinum. The S type thermocouple has good stability and high accuracy; therefore, it is used in those applications where high temperature needs to be measured with very good accuracy, such as in pharmaceutical industries, biotech industries, etc.

2.5.7.1.2.2.5 Type R thermocouples Like type S thermocouples, type R thermocouples are also made from platinum and rhodium. However, in R type thermocouples the positive lead is composed of 87 percent platinum and 13 percent rhodium and negative lead is composed of only platinum. Compared to the type S thermocouple, the range and the stability of the R type thermocouple is more. The temperature range of the R type thermocouple is between 0 and $1,600$ °C.

It is important to note that type J, K, and T thermocouples are the most common types of thermocouples. Type R and S thermocouples 66 ºC are used in high-temperature applications

2.5.7.1.2.3 APPLICATIONS OF THERMOCOUPLES

Depending on the temperature range and the applications, thermocouples can be selected:

- Type K thermocouples are used for measuring the temperature of process plants such as chemical production plants, petroleum refineries, etc.
- Type J is used for monitoring the temperature in a vacuum and inert metals.

- Type S or type R thermocouples are used to measure the temperature of steel and iron industries.
- In chemical production plants and petroleum refineries, a number of thermocouples are used to measure and monitor the temperature of the plant at different stages.
- Thermocouples are also used in thermostats as temperature sensors to measure the temperature of the office, showrooms, and homes.

2.5.7.1.3 Resistance temperature detector (RTD)

A resistance temperature detector (RTD) is a temperature sensor used for temperature measurements. It is an electronic device where, with the change in the temperature, the resistance of the device also changes. It has greater accuracy, stability, and repeatability. The RTD is an ideal temperature sensor for those applications that require very accurate temperature measurement. Further, it has good linear characteristics over a wide range of temperatures. This is because the resistance of the RTD varies linearly with the temperature. As the temperature of an object increases or decreases correspondingly, the resistance also changes accordingly.

2.5.7.1.3.1 WORKING PRINCIPLE OF RTD

We know that at a particular temperature, every metal has a definite value of resistance. Further, in RTD the value of resistance changes with the change in its temperature. Hence, we can calculate the temperature of metals by knowing their resistance. The resistance of an RTD at any temperature (t) can be calculated from the following formula:

$$R_t = R_0 \left(1 + \alpha (t - t_0) + \beta (t - t_0)^2 + ... \right) \tag{2.5}$$

Here, R_t and R_0 represent the resistance values at temperatures t and t_0 receptively, and α and β denote the material constants.

Generally, the RTD is constructed using copper, nickel, and platinum. In each of these metals the resistance varies differently with the temperature variations and they have different resistance- temperature characteristics. Depending on the material selection, the values of α and β will be different.

In the RTD, the value of change in resistance is also measured using a wheat stone bridge similar to a thermistor. A constant electric current is supplied to the bridge and corresponding to that there causes a voltage drop which results in a change in the electrical resistance of RTD. The RTD resistance can be calculated by measuring the voltage (V) across the two ends as shown in Figure 2.7. The relation between V and the RTD resistance can be given as:

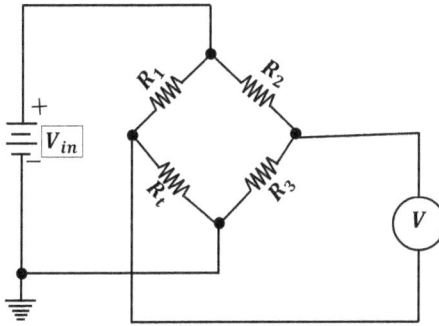

Figure 2.7 Temperature measurement using RTD in terms of voltage

$$V = V_{in}\left(\frac{R_tR_2 - R_1R_3}{(R_1 + R_t)(R_2 + R_3)}\right) \qquad (2.6)$$

Here, V_{in} represents input excitation voltage, resistances R_1, R_2, R_3 are selected depending on the accuracy needed by the application, R_t is the resistance of the RTD, and V denotes the output voltage that is measured from voltmeter.

Finally, the temperature of the object can be determined by converting the RTD resistance value using a calibration expression as given in Eqn. (2.6).

2.5.7.1.3.2 APPLICATIONS OF RTD

- In automotive, RTD is used to measure the engine temperature.
- RTD is used in communication and instrumentation for measuring the temperature of amplifiers, transistor gain stabilizers, etc.
- It is also used to measure the temperature of the washing machine, air conditioners, refrigerator, etc.
- Commonly used in industrial applications, including industrial boilers, petrochemical, exhaust gas monitoring, and food processing.

2.5.7.1.4 Bimetallic

A bimetallic strip is a device which is used for temperature measurements. It is based on the principle of thermal expansion where the volume of the metal changes with the change in temperature. Basically, it converts the temperature change into mechanical displacement using a bimetallic strip. The bimetallic strip consists of two different metals having different coefficients of thermal expansion that are bonded together. The bimetallic strip is usually constructed from steel and copper and, in some cases, is also designed from steel and brass. It works on the two very basic fundamentals of metals:

- Thermal expansion: depending on the temperature variation, metals either expand or contract.
- Temperature coefficient: at constant temperature metal expands or contracts differently.

2.5.7.1.4.1 WORKING PRINCIPLE OF BIMETALLIC STRIP

The bimetallic strip is formed from two thin strips of metals, usually steel and brass. These strips are joined together along the length to form a structure in which one end of the strip is fixed and the other end is left free, as shown in Figure 2.8 (a).

When the heat is applied to the bimetallic strip, its temperature rises. As a result, the strip will bend in the direction of a metal that has a lower temperature coefficient as illustrated in Figure 2.8 (b). Similarly, when the heat is removed, the temperature will decrease and the strip will expand in the direction of metal having a higher temperature coefficient, i.e., towards brass. The deflection of the strip indicates the temperature variation. To the dial of the thermometer, bending motion is connected and hence gives information about the temperature of the media.

2.5.7.1.4.2 APPLICATIONS OF BIMETALLIC STRIP

Bimetallic strips are commonly used in:

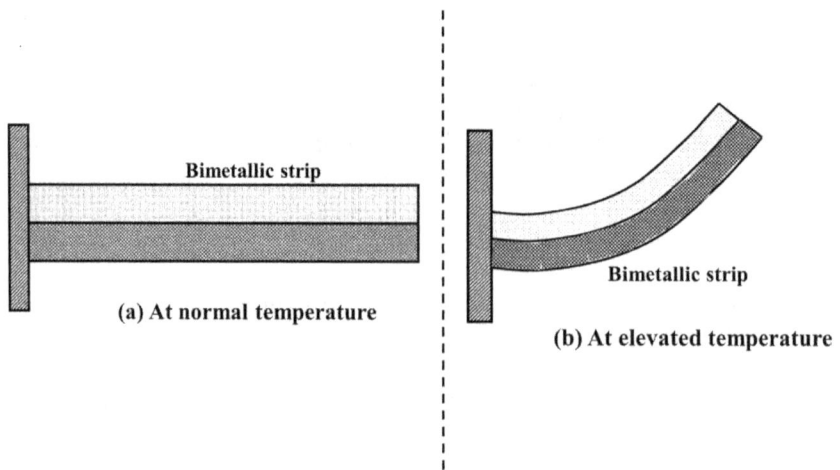

Figure 2.8 Bimetallic strip at two different temperatures (a) normal and (b) elevated

- Air conditioners.
- Thermostats.
- Control devices.
- Heaters.
- Ovens.
- Hotwires.
- Refineries.
- Oil burners.
- When a certain temperature is reached, the blade arches. Hence, it allows automatic switching of circuits to control heating and cooling of electrical devices [29].

2.5.8 Flow sensor

A flow sensor is a device used to measure the flow rate of a fluid. Flow sensors incorporate electrical, electronics, and also mechanical subsystems to measure any changes in the characteristics of a fluid. The fluid's flow rate is calculated from the measured physical properties. Flow rate measurement helps in the controlling of many industrial processes and also helps in the proper operation of machinery at an optimum performance level [30, 31]. Several flow sensors are available in the market, but here we will discuss only some famous types of flow sensors.

2.5.8.1 Differential pressure flow sensor

The differential flow meter is the most commonly used flow meter. It is used to measure the flow rate of gases, liquids, and stems. It is used in areas where high pressure, high temperature, and a large diameter play an important role.

2.5.8.1.1 Working principle

The differential flow meter works on the principle of creating an artificial constriction that causes obstruction in the flow of fluid as it passes through the constriction. This creates a pressure difference due to the pressure drop that occurs when the fluid flows through a constriction. This difference in the pressure is measured and is used to determine the flow rate. According to Bernoulli's principle, the pressure drop across the constriction is proportional to the square of the flow rate. The higher the pressure drops, the higher the flow rate.

Figure 2.9 illustrates that when a fluid flows through a constriction, it accelerates to a higher velocity (i.e., V_2) to conserve the mass flow, i.e., the mass of fluid flowing every second through a tube prior to the constriction must be equal to the fluid flowing through the tube at the constriction and,

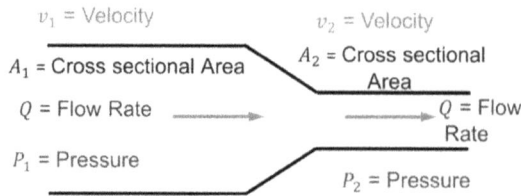

Figure 2.9 Flow rate measurement using differential pressure flow sensor by calculating the pressure difference

as a consequence of this, the fluid pressure drops. For a horizontal tube shown in Figure 2.9, V_1, P_1, and A_1 are the fluid velocity, pressure, and cross-sectional area of the tube prior to the constriction, respectively and V_2, P_2, and A_2 are the fluid velocity, pressure, and cross-sectional area of the tube at the constriction, respectively. Bernoulli's equation for the same can be written as [32]:

$$\frac{V_1^2}{2g} + \frac{P_1}{\rho g} = \frac{V_2^2}{2g} + \frac{P_2}{\rho g} \tag{2.7}$$

Since the fluid passing through the tube prior to constriction must be equal to fluid after the constriction, hence, we can write $\rho A_1 V_1 = \rho A_2 V_2$. But the flow rate "$Q$" of fluid passing through the tube per second is $A_1 V_1 = A_2 V_2$. Hence

$$Q = \frac{A_2}{\sqrt{1 - \left(\frac{A_2}{A_1}\right)^2}} \sqrt{\frac{2(P_1 - P_2)}{\rho}} \tag{2.8}$$

Hence, from Eqn. (2.8) it can be observed that the differential pressure generated is proportional to the square of the mass flow rate, Q, i.e., $Q \propto \sqrt{(P_1 - P_2)}$. Therefore, we can say that the measurements of the pressure difference can be used to measure the rate of fluid flow.

2.5.8.1.2 Types of differential pressure flow meter

The most common types of differential pressure meter are:

- Orifice plates.
- Venturi meter.
- Rotameter.

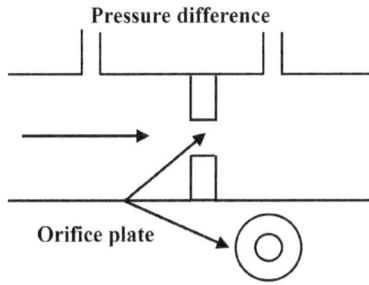

Figure 2.10 Flow rate measurement using orifice plate

2.5.8.1.2.1 ORIFICE PLATES

The orifice plate is used for the measurement and control of fluid flow. The shape of the orifice plate is very similar to a circular disc with a hole in the centre as shown in Figure 2.10. The disc is placed in the tube through which the fluid is flowing.

In flow control applications, the restriction in the path of the fluid is created using orifice plates. This restriction in the tube regulates the fluid flow and hence creates a pressure difference. The pressure difference is measured and is used to calculate the flow rate of the fluid. The orifice plate is cost-effective and used widely because of its simple design with no moving parts. The accuracy of the orifice plate is typically about 62 percent of its full range.

2.5.8.1.2.2 VENTURI METER

The venturi meter as shown in Figure 2.11 is a differential flow meter used to measure the flow rate of the fluid (gas or liquid) and is based on Bernoulli's theorem. Venturi tubes are a constriction in a fluid conduit and

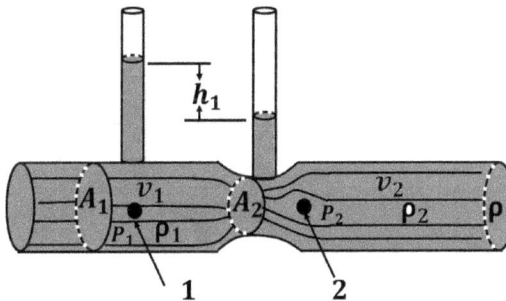

Figure 2.11 Measurement of flow rate using Venturi meter

when a fluid is passed through a constriction having a smaller cross-section, the static pressure decreases.

The Bernoulli equation gives the relationship between velocity and pressure can be defined as:

$$P_1 - P_2 = \frac{\rho}{2}\left(v_2^2 - v_1^2\right) \tag{2.9}$$

where P_1 and P_2 are pressure fluid pressure before and after the construction respectively, ρ is the fluid density and v_1 and v_2 are the fluid velocity before and after the constriction. It is important to note that the static pressure in the first measuring tube, i.e., "1" is higher than in the second measuring tube, i.e., "2". However, the fluid speed at "1" is lower than at "2," because the cross-sectional area at "1" is greater than at "2." Hence, by measuring the change in pressure, the flow rate of the fluid can be determined by using a venturi tube. The venturi meter is preferred in several industrial applications because it is simple to design and also a very accurate form of flow rate measurement.

2.5.8.1.2.3 ROTAMETER

A rotameter is also called a variable area flow meter and is used to measure the volumetric flow of liquids and gases. It has the simplest measuring technique where a liquid or gas is passed through a tapered tube that has a freely moving float as shown in Figure 2.12.

Floats are available in different shapes, like spheres, ellipsoids, etc. They also have a higher density than the fluid. Hence, it does not float in the absence of any fluid. When the liquid or gas passes through the tube the float rises because of a combination of the velocity head of the fluid and the buoyancy of the liquid. The falling and rising action of a float in a tapered tube provides a measure of flow rate. The float moves either up or down and comes to equilibrium when the weight of the float is equal to the upward force exerted by the fluid or gas. The float position can be related to the

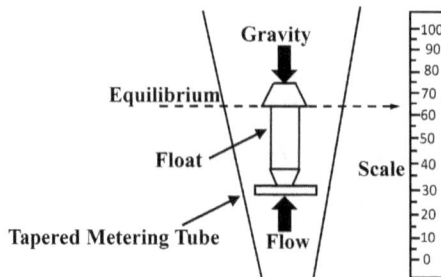

Figure 2.12 Flow rate measurement using rotameter by detecting the position of the float

volumetric flow rate. This position can be compared to a calibrated scale either placed next to the tube or through a scale on the tube itself.

2.5.8.1.3 Applications of flow sensor

- Flow sensors are used in the automobile industry to measure the air intake in the engine. This helps in delivering an accurate amount of fuel to the injector and hence provides the required fuel mixture to the engine.
- Flow sensors are also used in medical ventilators for delivering air and oxygen to the patients at a correct rate.
- Industrial applications require highly reliable air flowmeters for accurate monitoring of compressed air usage.
- Steam flow measurement.

2.5.9 Position sensor

The position sensor is a device used to detect an object's position such as doors, valves, etc. The sensed signal is converted into electrical signals which are suitable for processing, transmission, or control. These sensors are equipped with location tracking systems that help to determine the precise positions of work-in-progress, tools, and other production-relevant items within the facility [33, 34]. Figure 2.13 shows the different types of position sensors.

The position sensors can be contact type or non-contact type:

Contact sensors – these measure the position of an object through direct contact with the sensors. It is a device that uses a transducer to sense mechanical contact and gives an output signal when the measured object comes in contact with the sensor. The commonly used contact sensors are the potentiometer, strain gauge, etc. They are commonly used in robotics to

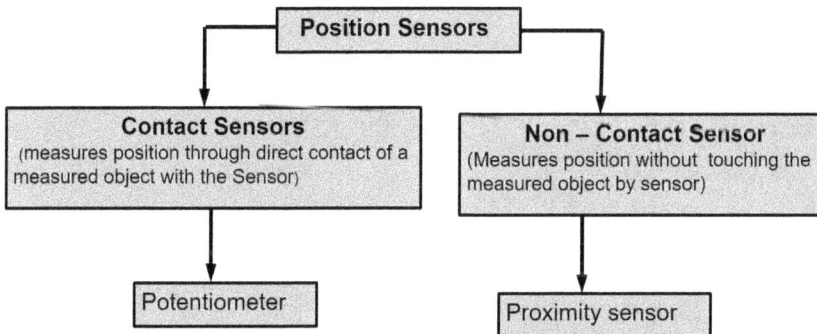

Figure 2.13 Block diagram representation of the position sensors

detect the change in position, velocity, acceleration or force/torque between the manipulator's joints and the end effector.

2.5.9.1 Potentiometer

A potentiometer is the simplest type of position sensor. It is a three terminal resistor having either sliding or rotating contact that forms an adjustable voltage divider. The resistance is varied manually to control the flow of electric current and hence, producing a voltage drop in accordance with Ohm's law. Potentiometers are passive devices and do not need any power supply or other electronics circuit in order to measure linear or rotary motion.

2.5.9.1.1 Working principle

Potentiometers work by varying the position of the sliding contact/wiper (b) over the uniform resistance "R" as shown in Figure 2.14. The potentiometer has two input terminals, namely "a" and "c," source voltage " V_{in} " that is applied across the entire resistance length, and the output voltage, representing the voltage drop across the fixed and sliding contact. To get the output voltage, the sliding contact is moved along the resistor and is calculated using Eqn. (2.10).

$$V_{out} = V_{in}\left(\frac{R_2}{R_1 + R_2}\right) \tag{2.10}$$

2.5.9.1.2 Types of potentiometers

Potentiometers are classified into two types: (a) linear potentiometer and (b) rotary potentiometer. The linear and rotary potentiometer consist of two fixed terminals, one moving terminal known as the wiper and a resistive strip known as a track. Figure 2.15 depicts a linear potentiometer with a straight track and rotary potentiometer having a circular track.

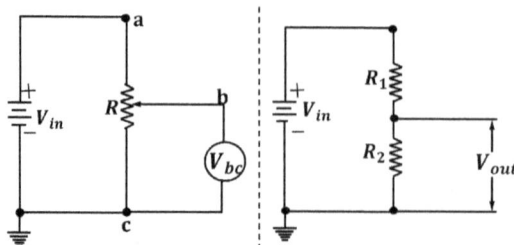

Figure 2.14 Schematic of the voltage divider circuit

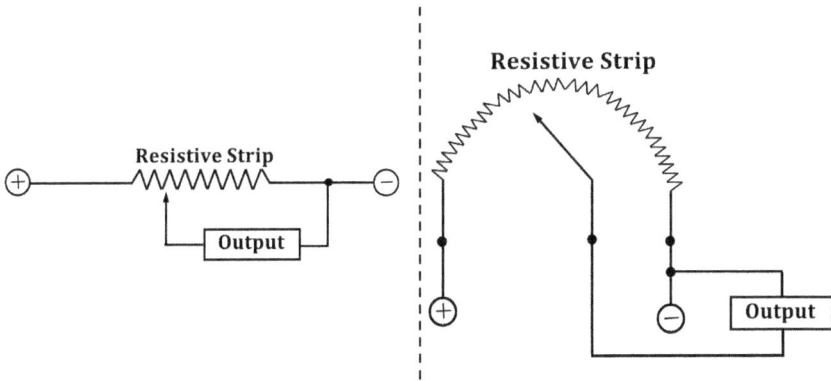

Figure 2.15 Two different types of potentiometer (a) linear and (b) rotary

2.5.9.1.3 Applications of potentiometer

Potentiometer are commonly used as

- Voltage dividers in the electronic circuit.
- Radio and television receivers for tone, volume and linearity control.
- In agricultural machinery, linear potentiometers are used for precise control of steering.
- Rotary position sensors are used within the steering systems of some marine vehicles, including submarines.
- Ticket barriers which require a ticket to be scanned before opening use rotary potentiometers to ensure the gate is open for the allocated amount of time. These are used on many ticket barriers, including the ones on the London Underground.

2.5.9.1.4 Proximity sensor

Proximity sensors have the ability to detect the nearby object without any physical touch and hence don't cause a scratch or damage to the object. Figure 2.16 represents the different types of proximity sensors.

2.5.9.1.4.1 INDUCTIVE PROXIMITY SENSOR

Inductive proximity sensors are non-contact electronic proximity sensors, which are used to detect metallic objects. The sensors operate under the electrical principle of inductance where a fluctuating current induces an electromotive force (EMF) when a target object comes in close proximity to the sensing area.

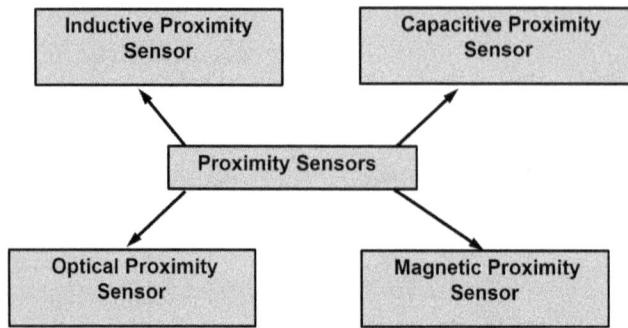

Figure 2.16 Block diagram presentation of the proximity sensor

2.5.9.1.4.2 CAPACITIVE PROXIMITY SENSOR

The capacitive proximity sensor is similar to the inductive proximity sensor. The main difference between the inductive and capacitive proximity sensor is that the latter produces an electrostatic field instead of a magnetic field. Hence, the sensing area of the capacitive proximity sensor is activated by both conductive and non-conductive materials, and thus is used to detect both metallic and non-metallic objects. A capacitive proximity sensor contains a high-frequency oscillating circuit along with a sensing surface formatted by two metal plates. When an object or target material comes into the sensing areas, the electrostatic field of the metal plates gets disturbed and causes a change in the capacitance of the proximity sensor. This change in the capacitance of the sensors is used to detect objects.

2.5.9.1.4.3 OPTICAL PROXIMITY SENSOR

Optical proximity sensors are used to detect almost all objects. They are expensive compared to inductive proximity sensors. The optical proximity sensors use the principle of triangulation of reflected infrared or visible light to measure the target objects. Here, when the target object comes into the sensor's range, the transmitted light from the sensor is reflected back to the sensor from the object and hence, detects the presence of an object.

The optical proximity sensors are extensively used in automated systems. These sensors are commonly known as light beam sensors of the thru-beam type or of the retro reflective type.

2.5.9.1.4.4 MAGNETIC PROXIMITY SENSOR

The non-contact magnetic proximity sensors are used to detect magnetic objects like permanent magnets. It uses GMR, i.e., giant magneto resistive technology to detect objects. Here the measuring unit consists of resistors along with several ferromagnetic or nonmagnetic layers. The two GMR resistors form a conventional Wheatstone bridge circuit which amplifies an output voltage signal that is related to the magnetic field strength impinging on the detector. Magnetic proximity sensors are more reliable than the simple position determining mechanical switches because they have no moving parts for wear or jam.

2.5.10 Smart sensor

With the development of many advanced technologies, smart sensors are also in the spotlight for industrial and other application areas. Ordinary sensors have been transformed into smart sensors with the integration of computing and IoT. These sensors have the ability to perform complex calculations from the collected data [35, 36]. In addition to greater flexibility, the smart sensors are also available in small sizes and are extremely flexible, hence converting bulky machines into highly intelligent technology. In addition to ordinary sensors, smart sensors are also equipped with intelligent/embedded algorithms, good signal conditioning, and digital interface capabilities. These make smart sensors a device with self-detection and self-correctness/awareness capabilities [37]. In IoT, smart sensors play a very important role. They help in converting real-time data into digital information which can be transmitted to a gateway [38]. Hence, smart sensors can easily predict and monitor the real time situation and can also take corrective actions instantly, depending on the applications requirement. In addition to this, smart sensors can also perform multi-layered complex tasks like motion detection, collecting raw data, analysis, filtering, adjusting amplitude, sensitivity, and communication, which are the major functions of the smart sensors [39]. The very common applications of smart sensors are wireless sensor networks (WSNs) where nodes are connected with multiple sensors and form a communication technology. Figure 2.17 (a) illustrates the building block of the smart sensors [40], and Figure 2.17 (b) describes the important features of smart sensors [37, 41].

2.5.10.1 Self-calibration capability

The sensor's calibration capability is defined as the ability of the sensor to determine the normal function of the sensor [42]. For different types of sensors, different calibration techniques are available:

- The calibration of sensors that give electrical output are done using known reference voltage.
- In the case of load cells that are used for weighing systems, the calibration is done by adjusting the output to zero under no load condition [43].

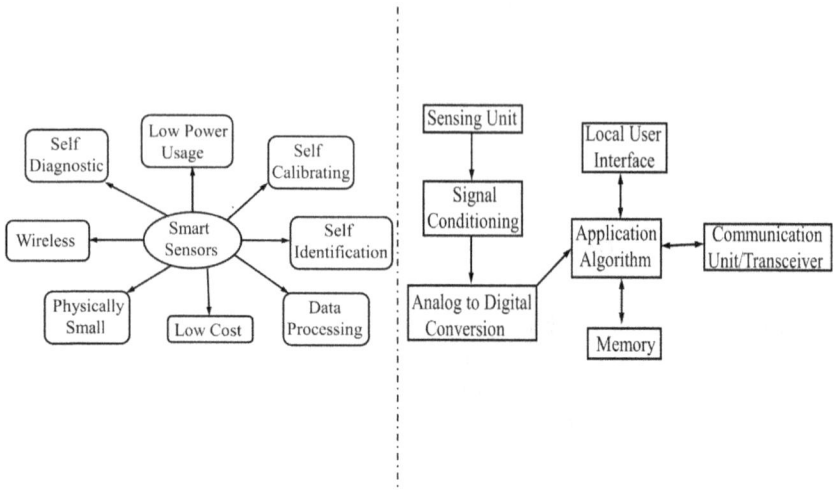

Figure 2.17 (a) Building blocks of smart sensor and (b) features of smart sensors

- Some sensors also use look-up tables for the calibration. However, the calibration using look-up tables requires a huge amount of memory capacity to store the correction points.

2.5.10.2 Self-diagnosis of faults

Smart sensors can also self-diagnose the fault, by monitoring the internal signals that are used for the indication of the defects. A major problem in self-diagnosis is differentiating between normal measurement deviations and a sensor fault [37]. To overcome this problem, several smart sensors first load numerous measured values around a set point and then calculate the minimum and maximum expected values for the measured quantity [44]. Further, to measure the effect of fault on the measured data, indeterminate techniques are used. This enables the continuous working of the sensor even after the fault has developed.

2.5.11 Nuclear sensors

Nuclear sensors are very rarely used due mainly to two reasons. First, they are expensive and second, these sensors are prone to being contaminated by background radiation; hence, they have to follow strict safety regulations for their use. Nuclear sensors are based on the principles of optical sensors.

2.5.12 Micro-sensors (MEMS sensors)

Micro-sensors are small transducers that can convert mechanical signals from energy sources to electrical signals. The typical sizes of these sensors range from 0.01 mm or 10^{-5} m to 5 mm. The micro-sensors are commonly used in industry to measure pressure, force, temperature, speed, etc. These sensors are standard in medical equipment like the sphygmomanometer, which measures blood pressure [45]. The main advantages of micro-sensors are that they are compact, economical, reliable, and have better efficiency.

2.5.13 Nano-sensors (NEMS)

The most recent development in sensing technology are nano-sensors. These sensors are based on nanotechnology and the size varies from 1 to 1,000 mm [37, 44]. These sensors are a part of the nano-electromechanical system (NEMS) and are used as accelerometers, biological sensors, nano-actuators, etc. The sensor has a lower production cost, smaller size, and reduced power consumption.

2.5.14 Actuator

An actuator is a part of a mechanical device or machine that converts the supplied energy into physical movements. Basically, an actuator needs an energy source for its operation. The source of energy can be electric current, hydraulic fluid pressure, or pneumatic pressure. The most commonly used actuators are (a) hydraulic actuators, (b) pneumatic actuators, (c) electric actuators, and (d) piezoelectric actuators.

2.5.14.1 Hydraulic actuator

A hydraulic actuator consists of a hollow cylindrical tube along with a piston that can slide. It uses hydraulic power to perform mechanical operation. An electric motor drives the pump and sends the fluid from a reservoir through control valves to the opposite sides of a cylinder. The term single acting is used when fluid pressure is applied to only one side of the piston. These cause the piston to move only in one direction. A spring is being used to bring the piston back to its original position or to give the piston a return stroke. The pressure can also be applied to both sides of the piston, known as double acting. The pressure difference created between the two sides of the piston causes the piston to move on either side.

2.5.14.2 Pneumatic actuator

A pneumatic actuator converts energy from high pressure compressed air or vacuum to rotary or linear motion. A pneumatic system or circuit works from the stored air pressure. It is generally available in two forms: diaphragm and piston operated. The air pressure built from the compressor is applied against the diaphragm or piston. This causes the movement of the valve actuator in the system and hence, results in mechanical motion.

Pneumatic actuators are very reliable and require much less maintenance and therefore are commonly used on production lines.

2.5.14.3 Electric actuator

An electric actuator is a device that converts electrical energy into linear or rotary motion. It uses an electric motor to generate the force that is necessary to create the required action or movement. The linear actuators vary the power in a straight-line fashion and are normally used for two tasks, namely push and pull. All linear actuators are not power driven, some linear actuators function manually using a rotating handle or hand wheel. Similarly, the rotary actuators are used to provide the rotary motion and are commonly used to control the valves like balls, butterflies, etc. Depending on the applications, actuators are available in various shapes and sizes.

2.5.14.4 Piezoelectric actuator

Piezoelectric actuators are transducers that are based on the piezoelectric effect which converts electrical energy into mechanical displacement or stress. The reverse is also possible using the inverse piezoelectric effect where applied mechanical displacement or stress produces an electrical charge. The piezoelectric actuators are used as high-precision positioning mechanisms; hence, they can control small mechanical displacement at high speeds. Piezoelectric actuators are commonly made using quartz, ceramics and lead zirconate titanate (PZT).

2.5.14.5 Smart actuator

Smart actuators are defined as the actuator integrated with the motor, controller, sensors, and communication systems. Compared to the above discussed conventional actuators, smart actuators make the system compact and economical. This is because the system using smart actuators eliminates the need for additional sensors and hence, saves the cost as well as space. Smart materials like shape memory alloy (SMA), piezoelectric, etc. based actuators are commonly used in robotics and medical applications. An SMA based self-sensing actuator has been designed and developed for robotic applications.

2.6 ROLE OF SMART SENSORS AND ACTUATORS

The main elements of Industry 4.0 are the industrial Internet of Things which mainly includes sensors and actuators. The sensors and actuators are used in industry for controlling and monitoring the different process variables [45, 46]. The role of smart sensors is already discussed in section 2.5.7. In Industry 4.0 smart sensors and actuators are the first level of the automation process. The different levels of automation structure are as follows [47]

Level 1: At this level, first sensors and actuators receive raw data from the manufacturing process and start controlling the process by monitoring the data.

Level 2: This level is an interfacing level that occurs between process and control parameters with the help of various input-output modules.

Level 3: Next comes the controlling stage, which helps in controlling the production process.

Level 4: The monitoring level monitors the whole manufacturing process and also includes a process management system.

Level 5: This level consists of detailed operational planning, quality specification management and data collection over the cloud.

2.7 CONCLUSION

The smart factory is defined as the factory where various Industry 4.0 technologies like digital technology, smart computing, IoT, etc. are combined with the production process and its operation.

In this chapter, Industry 4.0 and smart factories are described along with their characteristics and benefits. Different types of sensors technology and smart actuators that are commonly used in manufacturing and smart industries are discussed. As sensors and actuators are the most important components of the smart factory, their types, work, applications, advantages, and disadvantage are presented. The main differences between traditional and smart factory are discussed. From the discussion it can be concluded that smart factories are the automated flexible manufacturing system that can self-analyse the performance of the system and, if required, subsequently correct the process. They can adapt to or learn from new environments and can run the entire production processes autonomously and smoothly. Further it can be noted that in Industry 4.0 all the machines required for the manufacturing process are integrated with different devices. Each device has a unique IP address, allowing them to communicate with other IoT devices and helping in exchanging the large amount of data. This information received from the devices is optimized to improve the performance of the system.

In conclusion, this chapter focuses on the new phase of the industrial revolution that makes a system more flexible and gives many new directions for the manufacturing process.

REFERENCES

1. Lass, S., Gronau, N. A factory operating system for extending existing factories to industry 4.0. *Comput. Ind.* 2020;115:103128. doi: 10.1016/j.compind.2019.103128.
2. Shi, Z., Xie, Y., Xue, W., Chen, Y., Fu, L., Xu, X. Smart factory in industry 4.0. *Syst. Res. Behav. Sci.* 2020;37:607–617. doi: 10.1002/sres.2704.
3. Chang, W., Ellinger, A.E., Kim, K., Franke, G.R. Supply chain integration and firm financial performance: A meta-analysis of positional advantage mediation and moderating factors. *Eur. Manag. J.* 2016;34:282–295. doi: 10.1016/j.emj.2015.11.008.
4. Xie, Y., Yin, Y., Xue, W., Shi, H., Chong, D. Intelligent supply chain performance measurement in industry 4.0. *Syst. Res. Behav. Sci.* 2020;37:711–718. doi: 10.1002/sres.2712.
5. Ortiz, A.M., Hussein, D., Park, S., Han, S.N., Crespi, N. The cluster between internet of things and social networks: Review and research challenges. *IEEE Internet Things J.* 2014;1:206–215. doi: 10.1109/JIOT.2014.2318835.
6. Agostini, L., Filippini, R. Organizational and managerial challenges in the path toward industry 4.0. *Eur. J. Innov. Manag.* 2019;22:406–421. doi: 10.1108/EJIM-02-2018-0030.
7. Tortorella, G.L., Giglio, R., Van Dun, D.H. Industry 4.0 adoption as a moderator of the impact of lean production practices on operational performance improvement. *Int. J. Oper. Prod. Manag.* 2019;39:860–886. doi: 10.1108/IJOPM-01-2019-0005.
8. Cimini, C., Pirola, F., Pinto, R., Cavalieri, S. A human-in-the-loop manufacturing control architecture for the next generation of production systems. *J. Manuf. Syst.* 2020;54:258–271. doi: 10.1016/j.jmsy.2020.01.002.
9. Tao, F., Qi, Q., Liu, A., Kusiak, A. Data-driven smart manufacturing. *J. Manuf. Syst.* 2018;48:157–169. doi: 10.1016/j.jmsy.2018.01.006.
10. Wilkesmann, M., Wilkesmann, U. Industry 4.0—Organizing routines or innovations? *VINE J. Inf. Knowl. Manag. Syst.* 2018;48:238–254. doi: 10.1108/VJIKMS-04-2017-0019.
11. Shrouf, F., Ordieres, J., Miragliotta, G. Smart factories in industry 4.0: A review of the concept and of energy management approached in production based on the internet of things paradigm. In *Proceedings of the 2014 IEEE International Conference on Industrial Engineering and Engineering Management*, Bandar Sunway, Malaysia. 9–12 December 2014:697–701.
12. Kaur, J., Kaur, K. A fuzzy approach for an IoT-based automated employee performance appraisal. *Comput. Mater. Contin.* 2017;53:24–38.
13. Zanella, A., Bui, N., Castellani, A., Vangelista, L., Zorzi, M. Internet of things for smart cities. *IEEE Internet Things J.* 2014;1:22–32. doi: 10.1109/JIOT.2014.2306328.
14. Ghasemaghaei, M., Ebrahimi, S., Hassanein, K. Data analytics competency for improving firm decision making performance. *J. Strat. Inf. Syst.* 2018;27:101–113. doi: 10.1016/j.jsis.2017.10.001.
15. Gassmann, O., Kottmann, J. Technologie management in der Sensorik. *Wissensmanagement* 2002;8:19–24.

16. Benitez, G.B., Ayala, N.F., Frank, A.G. Industry 4.0 innovation ecosystems: An evolutionary perspective on value cocreation. *Int. J. Prod. Econ.* 2020;228:107735.

17. Choy, J.L.C., Wu, J., Long, C., Lin, Y.B. Ubiquitous and low power vehicles speed monitoring for intelligent transport systems. *IEEE Sens. J.* 2020;20:5656–5665.

18. Farhangi, H. Smart grid. In: Martin A. Abraham (Ed.) *Encyclopedia of Sustainable Technologies*, 2017, Pages 195–203, ISBN 9780128047927, https://doi.org/10.1016/B978-0-12-409548-9.10135-6.

19. Kimani, K., Oduol, V., Langat, K. Cyber security challenges for IoT-based smart grid networks. *Int. J. Crit. Infrastruct. Prot.* 2019;25:36–49.

20. Frank, A.G., Dalenogare, L.S., Ayala, N.F. Industry 4.0 technologies: Implementation patterns in manufacturing companies. *Int. J. Prod. Econ.* 2019;210:15–26.

21. Gattullo, M., Scurati, G.W., Fiorentino, M., Uva, A.E., Ferrise, F., Bordegoni, M. Towards augmented reality manuals for industry 4.0: A methodology. *Robot. Comput. Integr. Manuf.* 2019;56:276–286.

22. Punithavathi, P., Geetha, S., Karuppiah, M., Islam, S.H., Hassan, M.M., Choo, K.-K.R. A lightweight machine learning-based authentication framework for smart IoT devices. *Inf. Sci.* 2019;484:255–268.

23. Ivanov, D., Dolgui, A., Sokolov, B., Werner, F., Ivanova, M. A dynamic model and an algorithm for short-term supply chain scheduling in the smart factory industry 4.0. *Int. J. Prod. Res.* 2016;54:386–402.

24. Guideline sensors for industry 4.0. Available from: https://industrie40.vdma .org/en/viewer/-/v2article/render/25266556.

25. Bibby, L., Dehe, B. Defining and assessing industry 4.0 maturity levels-case of the defence sector. *Prod. Plan. Control* 2018;29:1030–1043.

26. Herrojo, C., Paredes, F., Mata-Contreras, J., Martín, F. Chipless-RFID: A review and recent developments. *Sensors* 2019;19:3385.

27. Landaluce, H., Arjona, L., Perallos, A., Falcone, F., Angulo, I., Muralter, F. A review of IoT sensing applications and challenges using RFID and wireless sensor networks. *Sensors* 2020;20:2495.

28. Jeon, B., Yoon, J.S., Um, J., Suh, S.H. The architecture development of industry 4.0 compliant smart machine tool system (SMTS). *J. Intell. Manuf.* 2020;31:1837–1859.

29. Howard, E.R. Thermostatic bimetal. *Eng. Sci.* 1942;5(4):16–24.

30. Zang, H., Zhang, X., Zhu, B., Fatikow, S. Recent advances in non-contact force sensors used for micro/nano manipulation. *Sens. Actuators A Phys.* 2019;296:155–177.

31. Palmer, K., Kratz, H., Nguyen, H., Thornell, G. A highly integratable silicon thermal gas flow sensor. *J. Micromech. Microeng.* 2012;22:65015.

32. Bolton, William. *Mechatronics: a multidisciplinary approach*. Vol. 10. Pearson Education, 2008.

33. Islam, M.M.M., Sohaib, M., Kim, J., Kim, J.-M. Crack classification of a pressure vessel using feature selection and deep learning methods. *Sensors* 2018;18:4379.

34. Petrov, R.V., Sokolov, O.V., Bichurin, M.I., Petrova, A.R., Bozhkov, S., Milenov, I., Bozhkov, P. Strength of multiferroic layered structures in position sensor structures. *IOP Conf. Ser. Mater. Sci. Eng.* 2020;939:012058.
35. Bibby, L.;Dehe, B. Defining and assessing industry 4.0 maturity levels-case of the defence sector. *Prod. Plan.Control* 2018;29:1030–1043.
36. Landaluce, H., Arjona, L., Perallos, A., Falcone, F., Angulo, I., Muralter, F. A review of IoT sensing applications and challenges using RFID and wireless sensor networks. *Sensors* 2020;20:2495.
37. Kim, J., Cho, H., Han, S.-I., Han, A., Han, K.-H. A disposable microfluidic flow sensor with a reusable sensing substrate. *Sens. Actuators B Chem.* 2019;288:147–154.
38. Dahlin, A.B. Size matters: Problems and advantages associated with highly miniaturized sensors. *Sensors* 2012;12:3018–3036.
39. Mulloni, V., Donelli, M. Chipless RFID sensors for the internet of things: Challenges and opportunities. *Sensors* 2020;20:2135.
40. Le, D.N., Le Tuan, L., Tuan, M.N.D. Smart-building management system: An Internet-of-Things (IoT) application business model in Vietnam. *Technol. Forecast. Soc. Chang.* 2019;141:22–35.
41. Zuo, G., Dou, Y., Chang, X., Chen, Y., Ma, C. Design and performance analysis of a multilayer sea ice temperature sensor used in polar region. *Sensors* 2018;18:4467.
42. Jeon, B., Yoon, J.S., Um, J., Suh, S.H. The architecture development of industry 4.0 compliant smart machine tool system (SMTS). *J. Intell. Manuf.* 2020;31:1837–1859.
43. Gattullo, M., Scurati, G.W., Fiorentino, M., Uva, A.E., Ferrise, F., Bordegoni, M. Towards augmented reality manuals for industry 4.0: A methodology. *Robot. Comput. Integr. Manuf.* 2019;56:276–286.
44. Shkel, M. Smart MEMS: Micro-structures with error-suppression and self-calibration control capabilities. *Proc. Am. Control Conf.* 2001;2:1208–1213.
45. Gupta, D., de Albuquerque, V.H.C., Khanna, A., Mehta, P.L. *Smart Sensors for Industrial Internet of Things*, Springer International Publishing, 2021.
46. Tortorella, G.L., Giglio, R., Van Dun, D.H. Industry 4.0 adoption as a moderator of the impact of lean production practices on operational performance improvement. *Int. J. Oper. Prod. Manag.* 2019;39:860–886.
47. Leng, J., Ruan, G., Jiang, P., Xu, K., Liu, Q., Zhou, X., Liu, C. Blockchain-empowered sustainable manufacturing and product lifecycle management in industry 4.0: A survey. *Renew. Sustain. Energy Rev.* 2020;132:110112.

Chapter 3

Industry 4.0 and 5.0 towards enhanced productivity and competitiveness

Ashok G. Matani

3.1 INTRODUCTION: INDUSTRIAL REVOLUTION TILL INDUSTRY 4.0

The history of industrial revolution for over the last two centuries has turned machine intelligence progressively.

- Industry 1.0/Mechanization (~1780 to 1870) – The first industrial revolution started in the mid-1780s following the introduction of mechanical production facilities with the help of water and steam power.
- Industry 2.0/Mass production (~1870 to 1970) – The second industrial revolution started around 1870 with the introduction of electrically powered mass production based on the division of labor.
- Industry 3.0/Automation (~1970 to 2010) – The third industrial revolution started in the early 1970s when the first programmable logistic controller (PLC) Modicon 084 was built. It enabled production automation through the use of electronics and IT systems.
- Industry 4.0/Smart factories (~2011 to ?) – The industrial revolution 4.0 gave birth to smart factories through the use of cyber-physical systems. It means that physical systems such as machines and robotics are being controlled by automation systems equipped with machine learning algorithms. Minimal input is expected from human operators. The development of new technologies like the Internet of Things (IOT), additive manufacturing (3D printing), robotics, and artificial intelligence has been a primary driver of the movement to Industry 4.0. These technologies enabled areas such as manufacturing execution systems, shop floor control, and product life cycle management to unleash their full potential [24].

DOI: 10.1201/9781003246466-3

3.2 SIGNIFICANCE/NEED OF INDUSTRY 4.0 AND INDUSTRY 5.0

Industry 4.0 is a confluence of trends and technologies that promise to reshape the way things are made. It is fueled by connectivity and spans a little more than cyber-physical systems, the Internet of Things, virtual reality, machine learning, big data, cloud computing, etc. Society 5.0 is a human-centered society that balances economic advancement with the resolution of social problems by a system that highly integrates cyberspace and physical space; such as smart homes, wearables, autonomous mobility, digital assistants, smart/renewable energy, etc. I encourage you to search for these terms in Google images. These two will go hand in hand for the next decade, feeding into each other [25].

3.3 BASIC INDUSTRIAL INTERNET OF THINGS (IIOT) CONCEPTS

There are foundational words and phrases to know before we decide whether we want to invest in Industry 4.0 solutions for our business:

* **Artificial intelligence (AI):** Artificial intelligence is a concept that refers to a computer's ability to perform tasks and make decisions that would historically require some level of human intelligence [3].
* **Cyber-physical systems (CPS):** Cyber-physical systems, also sometimes known as cyber manufacturing, refers to an Industry 4.0-enabled manufacturing environment that offers real-time data collection, analysis, and transparency across every aspect of a manufacturing operation [4, 6].
* **Cloud computing:** Cloud computing refers to the practice of using interconnected remote servers hosted on the Internet to store, manage, and process information.
* **Digitization:** Digitization refers to the process of collecting and converting different types of information into a digital format [7].
* **Ecosystem:** An ecosystem, in terms of manufacturing, refers to the potential connectedness of your entire operation – inventory and planning, financials, customer relationships, supply chain management, and manufacturing execution [8].
* **Enterprise resource planning (ERP):** Business process management tools that can be used to manage information across an organization.
* **IoT:** IoT stands for Internet of Things, a concept that refers to connections between physical objects like sensors or machines and the Internet.
* **IIoT:** IIoT stands for the Industrial Internet of Things, a concept that refers to the connections between people, data, and machines as they relate to manufacturing.

** **M2M:** This stands for machine-to-machine, and refers to the communication that happens between two separate machines through wireless or wired networks.
** **Machine learning:** Machine learning refers to the ability that computers have to learn and improve on their own through artificial intelligence – without being explicitly told or programmed to do so.
** **Real-time data processing:** Real-time data processing refers to the abilities of computer systems and machines to continuously and automatically process data and provide real-time or near-time outputs and insights.
** **Smart factory:** A smart factory is one that invests in and leverages Industry 4.0 technology, solutions, and approaches [9].

3.4 REQUIREMENTS OF 5.0

Better trained employees: With the technologies so advanced, trained personnel specializing in the field are required. Employees who know about robotics, AI, and also know the machines and operators well are required. Training of employees has also taken a big leap using the facility of education online via courses or any other means [10].
Perfect technology: For better efficiency and collaborative working, new and better technology is always going to be a need. This is not only required for safety and a better-planned future but also a better factory 5.0 [11].

3.5 ADVANTAGES OF 5.0

** **Cost enhancement:** Financial decisions and conditions will be improved since humans and machines will be working together.
** **Greener solutions:** With the new technologies coming in, concerns regarding the environment are being given utmost priority [12].
** **Creative personalization:** Personalization is a key aspect of Industry 5.0. With machines still doing the monotonous tasks, human interaction with them opens the gates for individual personalization [13].

3.6 PREPAREDNESS FOR INDUSTRY 4.0 AND SOCIETY 5.0

Our profile for Industry 4.0 is determined based on the level of claims we can make:

** **Aware:** We know the 9+ key technologies and four principles of Industry 4.0 understand its impact on our business and also have some idea on how to use it for our growth. We are aware. We are not yet prepared to address the change [14].

** **Agile follower:** We know which technology to adopt when, and we will be able to do it as soon as our customers are ready for it. We have the necessary skill set to follow the industry leaders, in time, without sweating out of business [15].

** **Smart forecaster:** We understand the opportunity for us to enhance our customer's experience. We have embarked on adoption of some of the relevant technologies to create value proposition in anticipation of customer demand. Our leadership team can see the near future, and invest wisely. Our organization has a dependable and proven innovation process [16].

** **Visionary trendsetter:** We are pushing the boundaries in some of the top nine technologies and educating our customers through enhanced experience. We understand the risk in creating customer demand. Our organization has an exploratory mindset. We get it and intend to lead the consumers in this [17].

** **Robust and resilient:** We have the three things required to be robust and resilient as the fourth revolution unfolds. We can handle disruptions. We are already an influence on society 5.0

3.7 PREPARING FOR INDUSTRY 4.0 AND SOCIETY 5.0

To move from one profile to another, we need to progressively create a strategic road-map, and systematically build the expertise to create offerings suited for the fourth revolution. The +4π Innovation Framework is a proven set of best practices, structured to help businesses strengthen the culture of innovation to grow through the next industrial and social revolution [18].

a. **Innovation strategy:** Defines what products, services, and cyber-physical integration should a company develop with a timeline to survive and thrive over the next few years. A set of tools progressively adds rigor to the strategy through an in-depth evaluation of impact and opportunities in Industry 4.0 and eventually society 5.0 [19].

b. **Innovation capital:** Defines what skills, need for knowledge, and assets required and when, to support the innovation strategy. Set of tools guide Industry 4.0 mindset, skills, leadership, organization, and networking, which is different from what is currently practiced for most successful businesses [20].

c. **Innovative activity:** Defines the process to systematically and rapidly go from ideation to monetization. The desired speed of innovation demands increased focus on hackathons, burst innovation events, sustained open innovation, and business model redefinition, how this can change the face of your business and leave a lasting impression on society, if not humanity.

d. **Lean innovation:** Aligns products, processes, employees, customers, and business metrics, to continuously improve the efficiency and productivity of the enterprise. This set of tools guide the employee morale and customer engagement from management objectives, providing synergy benefits making the whole greater than the sum of individual pieces.

e. **Value from synergy:** A significant value contributed by the framework comes from sequence, dependence, and integration; where tools use a common set of parameters and indicators across the enterprise. This enables employees to break through the silos and ceilings, holding the company growth [21].

3.8 LIMITATIONS OF INDUSTRY 4.0 AND HIGHLIGHTS OF INDUSTRY 5.0

3.8.1 Industry 5.0 creates even higher-value jobs than Industry 4.0

Industry 4.0 focuses on the interconnectedness of machines and systems in order to achieve optimum performance to improve efficiencies and productivity. Industry 5.0 is touted as taking it a step further and refining the interaction between humans and machines. What we'll see is greater collaboration between the two: the ultra-fast precision of automated technology works with a human's critical thinking skills and creativity. The idea is that Industry 5.0 creates even higher-value jobs than Industry 4.0, because humans are taking back design responsibility, or work that requires creative thinking [22].

3.8.2 Industry 4.0 focuses on productivity using technology, Industry 5.0 adds human and sustainable approaches

Industry 4.0 describes how manufacturers will use technology to cope better in a changing world and economy. It is primarily a techno-economic vision, indicating how more general technological advancements, often originated in non-industrial contexts, will be brought to bear on industrial value chains and how they will change industry's economic position. Since its birth ten years ago, Industry 4.0 has been focusing on technology-driven methods for increasing the efficiency and productivity of different industries, taking less into consideration the principles of social fairness and sustainability. While enabling and deploying the latest technologies across most industries is paramount to stay competitive, focusing on innovation and optimizing industrial output is not enough. A narrow approach to industrial output and profit is becoming increasingly untenable and fails to account correctly for the environment and the costs for society.

Globalization, climate change, and the production and supply chain challenges we are experiencing now have taught us that technology alone is not the solution to address current world problems. Covid-19 has highlighted the need to reevaluate current working methods and approaches. We need to invest in automation, digitalization, and artificial intelligence, but not forgetting sustainability and the human factor. That's why the European Union started a new approach, based on the Industry 4.0 revolution, adding the concepts of circular economy, human-centric technology, sustainability, and resiliency [23].

Industry 5.0 is characterized by going beyond producing goods and services for profit. It shifts the focus from the shareholder value to stakeholder value and reinforces the role and the contribution of industry to society. It places the well-being of the worker at the center of the production process and uses new technologies to provide prosperity beyond jobs and growth while respecting the production limits of the planet. It complements the existing Industry 4.0 approach by specifically putting research and innovation at the service of the transition to a sustainable, human-centric, and resilient European industry.

3.9 THE EUROPEAN UNION HAS DEFINED LIST OF TECHNOLOGIES TO SUPPORT CONCEPT OF INDUSTRY 5.0

- Human-centric solutions and human-machine-interaction technologies that interconnect and combine the strengths of humans and machines.
- Bio-inspired technologies and smart materials that allow materials with embedded sensors and enhanced features while being recyclable.
- Real time-based digital twins and simulation to model entire systems.
- Cyber safe data transmission, storage, and analysis technologies that are able to handle data and system interoperability.
- Artificial Intelligence, e.g., to detect casualties in complex, dynamic systems, leading to actionable intelligence.
- Technologies for energy efficiency and trustworthy autonomy, as the technologies mentioned earlier, will require large amounts of energy.

Driverless cars with artificial intelligence (AI) and automated supermarkets run by collaborative robots (cobots) working without human supervision have sparked off new debates: what will be the impacts of extreme automation, turbocharged by the Internet of Things (IoT), AI, and the Industry 4.0, on big data and omics implementation science? The IoT builds on (1) broadband wireless internet connectivity, (2) miniaturized sensors embedded in animate and inanimate objects ranging from the house cat to the milk carton in your smart fridge, and (3) AI and cobots making sense of big data collected by sensors. Industry 4.0 is a high-tech strategy for manufacturing automation that employs the IoT, thus creating the smart factory.

Extreme automation until "everything is connected to everything else" poses, however, vulnerabilities that have been little considered to date [24].

First, highly integrated systems are vulnerable to systemic risks such as total network collapse in the event of failure of one of its parts, for example, by hacking or Internet viruses that can fully invade integrated systems [2]. Second, extreme connectivity creates new social and political power structures. If left unchecked, they might lead to authoritarian governance by one person in total control of network power, directly or through her/his connected surrogates. We propose Industry 5.0 that can democratize knowledge co-production from big data, building on the new concept of symmetrical innovation. Industry 5.0 utilizes IoT, but differs from predecessor automation systems by having three-dimensional (3D) symmetry in innovation ecosystem design: (1) a built-in safe exit strategy in case of demise of hyperconnected entrenched digital knowledge networks. Importantly, such safe exists are orthogonal – in that they allow "digital detox" by employing pathways unrelated/unaffected by automated networks, for example, electronic patient records versus material/article trails on vital medical information; (2) equal emphasis on both acceleration and deceleration of innovation if diminishing returns become apparent; and (3) next generation social science and humanities (SSH) research for global governance of emerging technologies: Post-ELSI Technology Evaluation Research' (PETER). Importantly, PETER considers the technology opportunity costs, ethics, ethics-of-ethics, framings (epistemology), independence, and reflexivity of SSH research in technology policymaking. Industry 5.0 is poised to harness extreme automation and big data with safety, innovative technology policy, and responsible implementation science, enabled by 3D symmetry in innovation ecosystem design [1].

Industry 5.0 is about building complex and hyperconnected digital networks without compromising long-term safety and sustainability of an innovation ecosystem and its constituents. Considering built-in orthogonal safe exits from the digital networks, recognizing the need for both acceleration/deceleration, and innovation in global governance for technology policy are three measures to bring about a 3D symmetry in future applications of Industry 5.0.

Industry 5.0 is poised to harness extreme automation and big data with safety, innovative technology policy, and responsible implementation science, enabled by 3D symmetry in innovation ecosystem design.

3.10 HOW IOT, BIG DATA, AND AI CAN ENABLE A SOCIAL TRANSFORMATION TO SOME OF THE FOCUS SECTORS

Let us see how IOT, big data and AI can enable a social transformation to some of the focus sectors:

3.10.1 Healthcare

- Connected healthcare and seamless share of information between medical data users.s
- Remote medical care and use AI and robots at nursing-care facilities to support people's independence.

3.10.2 Mobility

- Autonomous vehicles for public transportation.
- Improve distribution and logistics efficiency using drones.

3.10.3 Infrastructure

- Sensors, AI and robots for inspection and maintenance of roads, bridges, tunnels and dams.

3.10.4 FinTech

- Blockchain technology for money transfer.
- Open API to Fin tech firms and banks.
- Promote cashless payment.

3.11 EXAMPLES OF INDUSTRIAL INTERNET OF THINGS (IIOT) ALONG WITH THEIR APPLICATIONS

MAN: MAN is a truck and bus company. As the company provides its customers with the tracker which spots engine faults or other potential failures, it saves time and money for the customers.

Siemens: Siemens is a German multinational conglomerate company. The company wanted to build fully automated, Internet-based smart factories. The company builds automated machines for the likes of BMW. Siemens introduced an operating system called mind sphere, the cloud-based IoT unit from Siemens, which basically aggregates the data from all the different vital components of a factory and then processes them through rich analytics to produce useful results.

Caterpillar (CAT): CAT is an American machinery and equipment firm. The company uses augmented reality (AR) applications to operate machines from fuel levels to when air filters need replacing. The company sends basic instructions on how to replace it via an AR app. The company produces its industrial machinery with intelligent sensors and network capabilities which allows users to optimize and monitor processes closely. Caterpillar has brought about 45 percent efficiency into its production by putting IIoT to use. Tom Buckler, IoT, and Channel Solutions Director of Caterpillar said that the customer's satisfaction is at the forefront of their efforts. Caterpillar, in order to deliver real-time information to all the dealers regarding their equipment, joined hands with AT&T's IoT services in early 2018. With the help of AT&T, they have widespread connectivity of resources.

Airbus: This is a European multinational aerospace corporation. The company had launched a digital manufacturing initiative known as "Factory of the Future" to streamline operations and increase production capacity. The employees use a tablet or smart glasses (designed to reduce errors and bolster safety in the workplace) and smart devices to assess a task and communicate with the main infrastructure or locally with operators and then send that information to a robotic tool that completes it.

ABB: ABB is a Swiss-Swedish multinational corporation. The company is basically into the production of robots. It uses connected, low-cost sensors to monitor the maintenance of its robots to prompt repairs before the parts break. The company is using connected oil and gas production to solve hindrances at the plant thereby achieving business goals in a cost-effective way. The company had developed a compact sensor that is attached to the frame of low voltage induction motors, where no writing is needed. By using these sensors, the company gets the information about motor health via smartphones or on the Internet by a secure server.

Fanuc: Fanuc is one of the largest suppliers of industrial automation equipment in the world. The company developed the FIELD System

Table 3.1 Differences between Industry 4.0 and Industry 5.0

Industry 4.0	Industry 5.0
The aim is to do process automation.	The aim is for balancing machine-human interaction.
The technology was the most important.	Collaboration between humans and machines is the most important.
Completely virtual environment.	The transition back to the real environment.
Cut down the number of people as new smart technologies started being implemented.	Increased number of humans in touch with machinery.
Smarter and better-connected machines with the workspace.	Merging cognitive computing with human intelligence.
No personalization or customization is possible.	Personalization and customization are possible making every product better and made according to individual preferences.
It is still dangling between renewable and non-renewable sources.	It is more advantageous to nature since renewable sources will now be used more.

Table 3.2 Industrial Internet of Things (IIoT) and Internet of Things (IoT): How they differ

Basis	IoT	IIoT
Utility	It is mainly designated for individual customers, which can be used in homes and offices.	It is used in a commercial area that is industries.
Security	Security is not an issue in IoT in comparison to IIoT, as it does not include handling industrial processes.	Security is a major concern in IIoT as organizations and businesses at large scale.
Degree of application	It uses applications with low-risk impact.	It uses more sensitive and precise sensors.
Cost	It is less expensive as the technology of IoT has been introduced by various companies.	It is more expensive in comparison to IoT as it has sensitive devices and industrial applications.

(Fanuc Intelligent Edge Link & Drive System), an open platform that enables the execution of various IIoT applications that focus on heavy devices like robots, sensors, and machine tools. Alongside cloud-based analytics, Fanuc is utilizing sensors inside its robotics to anticipate any failure in the mechanism. With the help of this, the supervisors are able to keep up with the schedule and reduce costs.

Magna Steyr: This is an Austrian automotive manufacturer. The company offers production flexibility by using the concept of smart factories. The network system of the factory is digitally based. It is also using Bluetooth to test the concept of smart packaging and help the employees to better track the assets and efficiency between operations.

John Deere: This is an American corporation that manufactures agricultural, forestry, and construction machinery. The company bought into the self-driving vehicle revolution which no other company did. John Deere is the first company that bought into the concept of GPS in tractors. The company has deployed telematics technology for predictive maintenance applications.

Tesla: Tesla is an American automotive and energy company, specialized in the manufacturing of electric vehicles. The company is leveraging IT-driven data to move its business forward. It improves the functionality of the products via software updates. Autonomous Indoor Vehicles by Tesla have changed the way batteries are consumed. These batteries are chargeable on their own without any interruption. Tesla also introduced a feature that helped the customers to control and check the devices from anywhere through their smartphones.

Table 3.3 Two research approaches to situate industry 5.0 in society: Ethical, legal, and social implications (ELSI versus post-ELSI technology evaluation research)

Comparison	Ethical, legal, and social implications	Post-ELSI technology evaluation research
Aim	A "science enabler" position that seeks to make science proceed (or obtain legitimacy) through a narrow and linear innovation framework.	Critical social science research to situate new technology and science in a broader societal and political context.
	Tends to subscribe to technological determinism.	Questions the conceptual frames (epistemology) of scientific knowledge and innovation.
Analytical focus	Downstream impacts of science and new technology.	Upstream, anticipatory and design focus.
	Tends to view the technology future as preordained and imminent.	Views multiple possible technology futures (in plural).
Politics of technology	Usually bracketed out.	Critical social science research on politics and power relationships among innovation actors; "Unpacks the politics," thus making science and social science more accountable.
Analytical distance between the analyst and technology	The analyst is embedded among science and technology actors, or has a narrow analytical distance from science/technology, which may limit the ability of SSH scholars to say "No" to science and technology teams, thus risking co-option and endorsement of technological determinism.	The analyst tends to operate at a safe analytical distance from science, technology, and its actors. Greater empowerment of the SSH scholars for the ability to say "No" to science and technology teams and sustain independent critical SSH analysis.
Reflexivity and the role of human values in knowledge coproduction	Reflexivity, if considered at all, is limited to the context in scientific communities (Type I reflexivity).	Reflexivity is considered in both scientific communities and in SSH research. Considers politics and human values as factors that shape both science and technology (Type I reflexivity) and social knowledge and SSH scholars (Type 2 reflexivity).
Time frame	Post-hoc or post-facto scientific impacts.	Anticipatory or real time with technology development.
Opportunity costs considered	No.	Yes.
Symmetry in design	Accelerator role primarily.	Accelerator and decelerator for science and technology.
	Asymmetrical, and focused on enabling science through the lens of technological determinism, rather than the symmetrical accelerator/ decelerator function.	Has a built-in sustainability strategy for "safe exits" from entrenched and hyper connected networks in the digital age.

Hortilux: The company provides lighting solutions by introducing Hortisense, a digital solution that safeguards various operations. It uses smart sensors operated through the cloud to monitor the light levels and efficiency of the offered light. This information can be monitored and checked from anywhere on any device.

3.12 INDUSTRY 4.0 – LEGAL CHALLENGES AND STRATEGIES TO OVERCOME

Industry 4.0 (I4.0) is probably the most disruptive concept for most industries, affecting not only revenue and cost structures but also shaking up the core business and operating models. One aspect that is often neglected, but will have a major impact on the success of Industry 4.0, are legal issues and how they will affect digital transformation.

Pinsent Masons, a UK-based legal firm, identified in a study focusing on the preparedness of mid-sized German companies for Industry 4.0 and legal issues published in 2016, that 66 percent of companies see an increasing demand for legal support in the future. Although only 28 percent of these companies think that legal risks, for example related to data protection, will hamper the digital transformation of their company, major deficits exist in the identification of relevant legal risks.

Focus areas that will be influenced by implementing Industry 4.0 principles will be:

- Product liability, contractual liability, and distribution/assignment of risk.
- Data protection and IT security.
- Labor laws.
- Intellectual property.

3.12.1 Liability

It will be essential for any company that pursues the digital transformation of its business processes to carefully examine all of the legal challenges that it will face his will include risks that stem from the integration of external partners (e.g., R&D partners, suppliers and customers) in the supply chain of the company, with regard to data protection, security, and agreements on liability. The inclusion of respective insurances such as cyber insurance can avoid lengthy and costly lawsuits. Cyber insurances cover damages that will result from cyber-attacks or IT-system failures. In addition, the rework and/or adaptation of a company's general terms and conditions might also be necessary.

3.12.2 Data protection and IT security

Data protection and IT security have to be the responsibility of a company's top management in the advent of Industry 4.0. Apart from the organizational and technical impacts, the legal aspects of the digital transformation of processes and the introduction of new business models are to be taken into account from the beginning. The fact that personal data is protected will generate new challenges since data analytics will deliver a relationship with the individual. Hence each requires approval first. Complexity is also added by the fact that IT security and data automotive OEMs intending to collect data from their customers' cars will need their approval first. Complexity is also added by the fact that the various laws being faced in their markets will be complicated.

3.12.3 Labor laws

New technologies like artificial intelligence (AI) being introduced in the manufacturing process bring potential conflicts with existing labor laws and employee representative groups. This is understandable since the implementation of handling robots in assembly might make existing workers obsolete. In order to avoid risk in digital transformation processes from employees and their representatives, company management must involve them early on in the planning phase. Open communication and early information about planned digital transformation projects are key to obtaining buy-in. Industry 4.0 will enable increasingly flexible work time models. Management, employees, and their representative groups are asked to define agreements that allow for more. Digital transformation will also bring new job profiles and make existing ones and regulations concerning overtime obsolete. Therefore, the qualifications of the existing workforce towards the requirements of the new digital world will be crucial. Both management and flexible employees have to support and control this change.[5]

3.12.4 Intellectual property

Alongside this, Industry 4.0 will also have an impact on the protection of intellectual property. The company and its employees have to acknowledge that the legal protection of R&D, production, and company data is insufficient. Existing legal frameworks will not keep pace with the impact digitization will have on the protection of intellectual property. The expected increase in intellectual property disputes among partners and other parties can only be countered by the inclusion of it. Industry 4.0 will necessitate that companies pursuing digital transformation need to make substantial efforts to master legal changes in data protection, IT security, and labor law. There will need to be a general effort for the companies to safeguard their business from a legal point of view.

3.13 NEW REGULATORY CHALLENGES
AND RESEARCH DIRECTIONS

The convergence between Web 4.0, Industry 4.0, and the IoT (i) has already challenged the regulatory landscape, relating to law, governance, and the legal profession; (ii) brought about new regulatory challenges regarding, e.g., legal liability, data rights, data protection, trade restrictions, agreements, standards, contract models, supervision, surety, monitoring, and control; and (iii) created and stabilized new regulatory (or socio-legal) ecosystems that bind together all related stakeholders. The IoT is changing the social nature, function, and perspective of regulatory systems, both in their public and private dimensions.

Recent Gartner reports have highlighted that legacy silos of systems, data, and processes continue to limit government participation in broader digital ecosystems and constrain the implementation of fully digital end-to-end citizen services. These results also reflect the evolution of the web from Web 3.0 to Web 4.0, the emergence of Industry 4.0, and the construction of regulatory ecosystems. Open data can enable greater transparency, higher levels of citizen trust, better public service delivery, and more effective policy-making, but opening up data does not mean having to make it public.

Setting up platforms or apps for citizens' participation is not ensuring tangible results, it does not lead per se to reuse and value creation. Something else is needed, as the roles of citizens, consumers, stakeholders, and actors might be also changing in the new data-driven scenarios of the IoT. From 2018 on, Rules as Code, a regulatory movement fostered by some government agencies, civil servants, and entrepreneurs in New Zealand, Australia, Canada, France, and some other countries, will try to facilitate the enhancement of citizens' rights and a faster drafting and implementation of legal provisions by means of computer languages. Better rules and legislation as code are parallel developments as well (i) to design policies and (ii) to create and publish regulations, legislation, and policies as machine- and human readable.

3.14 CONCLUSIONS

As governments like Germany overhaul their economic strategies in the face of unprecedented challenges, including an exponentially faster rate of technological change, meaningful and relevant changes in education are urgently needed to achieve more inclusive and sustainable development for all, not just for the privileged few. To shorten the period of social pain and maximize the period of prosperity for all, education systems need to undergo transformative change. Rapid evolution of information and communications technology (ICT) is bringing drastic changes to society and industry. Technologies and IT infrastructure elements of business intelligence will impact the supply chain activities through cost-reduction opportunities and an increase of the process-transparency, therefore processes will be more digital and technological, where the company's personnel are able to acquire and share information using the business intelligence technology from anywhere around the globe. Procurement processes will be optimized, as suppliers can be fully flexible and autonomously chosen by specific software. Smartphone apps will have an impact on the organization of the supply chain activities from a technological perspective as well. In future, each employee will be equipped with this kind of mobile devices, interact with colleagues, perform time-management, and execute specific activities in the manufacturing process by using the smartphone and smartphone apps. Specific apps will be created to enhance the efficiency of the production processes – track and trace systems of specific product components, or by assisting software for the human activities in the company. The medical industry is a leading industry which is already including the smartphone apps within its supply chains. The term Industry 5.0 has been introduced to the research areas which are considered as the next industrial revolution which is a more systematic transformation that includes impact on civil society, governance, and structures, and human identity in addition to solely economic/manufacturing ramifications.

REFERENCES

1. A. Cardenas, S. Amin, S. Sastry (2008) The 28th international conference on distributed computing systems workshops, Beijing, 495–500.
2. A. Reiner (2014) Industry 4.0 - Advanced engineering of smart products and smart production. 19th International Seminar on High technology, Piracicaba, Brasil, 9th October, 2014.
3. S. Ali, S.B. Qaisar, H. Saeed, M.F. Khan, M. Naeem, A. Anpalagan (2015) Network challenges for cyber physical systems with tiny wireless devices: A case study on reliable pipeline condition monitoring. *Sensors (Basel)* 15(4):7172–205.
4. C.A. Valdez, P. Brauner, A.K. Schaar, A. Holzinger, M. Ziefle (2015) Reducing complexity with simplicity - Usability methods for industry 4.0. 19th Triennial Congress of the International Ergonomics Association, IEA 2015.
5. Dr. A.G. Matani (2001) Managing new product innovations. *Industrial Engineering Journal* 4(1):21–23.
6. Dr. A.G. Matani (1999) Strategies for better waste management in industrial estates. *Journal of Industrial Pollution Control* 22(1):67–72.
7. E. Qin, Y. Long, C. Zhang, L. Huang (2013) International conference on human interface and the management of information - HIMI 2013. Human interface and the management of information. *Information and Interaction for Health, Safety, Mobility and Complex Environments* 3(1):173–180.
8. F. Shrouf, J. Ordieres-Mere, A. García-Sánchez, M. Ortega-Mier (2014) Optimizing the production scheduling of a single machine to minimize total energy consumption costs. *Journal of Cleaner Production* 67:197–207.

9. Z.X. Guo, E.W.T. Ngai, C. Yang, X. Liang (2015) An RFID-based intelligent decision support system architecture for production monitoring and scheduling in a distributed manufacturing environment. *International Journal of Production Economics* 159:16–28.

10. H. Hartenstein, K. Laberteaux (2010) *VANET Vehicular Applications and Inter-Networking Technologies* (1st ed.). Wiley Publication, London.

11. H. Carvalho, V. Cruz-Machado (2011) Integrating lean, agile, resilience and green paradigms in supply chain management (LARG_SCM). Supply Chain Management, InTech.

12. S. Jeschke, C. Brecher, H. Song, D. B. Rawat (Eds.). (2017) *Industrial Internet of Things and Cyber Manufacturing Systems*. Springer International Publishing, 3–19.

13. J. Macaulay, L. Buckalew, G. Chung (2015) *Internet of Things in Logistics*. DHL Customer Solutions & Innovation.

14. J. Nasser (2014) Cyber physical systems in the context of industry 4.0. Automation, quality and testing, robotics, 2014 IEEE International Conference on IEEE.

15. K. Witkowski (2017) Internet of things, big data, industry 4.0 – Innovative solutions in logistics and supply chains management, science direct. *Procedia Engineering* 182:763–769.

16. L. Barreto, A. Amaral, T. Pereira (2017), Industry 4.0 implications in logistics: an overview, Procedia Manufacturing, Volume 13, Pages 1245–1252, ISSN 2351-9789, https://doi.org/10.1016/j.promfg.2017.09.045. (https://www.sciencedirect.com/science/article/pii/S2351978917306807).

17. L. Barreto, A. Amaral, T. Pereira (2017) Industry 4.0 implications in logistics: An overview. *Procedia Manufacturing* 13(1):1245–1252.

18. L. Heuser, Z. Nochta, N.C. Trunk (2008) *ICT Shaping the World: A Scientific View*. ETSI, Wiley Publication, London.

19. Y.Q. Lu, X. Xu (2017) A semantic web-based framework for service composition in a cloud manufacturing environment. *Journal of Manufacturing Systems* 42:69–81.

20. Valdeza, André Calero, et al. "Reducing complexity with simplicity-usability methods for industry 4.0. "*Proceedings 19th triennial congress of the IEA. Vol. 9.* 2015.

21. Macaulay, J., L. Buckalew, and G. Chung. "Internet of things in logistics. DHL." *Trend Research Cisco Consulting Services A collaborative report by DHL and Cisco on implications and use cases for the logistics industry* (2015).

22. R.Y. Zhong, X. Xu, E. Klotz, S.T. Newman (2017) Intelligent manufacturing in the context of industry 4.0: A review. *Engineering* 3:616–630.

23. S. Hasan, N. Siddique, S. Chakraborty (2013) *Intelligent Transport Systems. 802.11-based Roadside-to-Vehicle Communications*. Springer Publication, New York.

24. F. Shrouf, G. Miragliotta (2015) Energy management based on internet of things: Practices and framework for adoption in production management. *Journal of Cleaner Production*: Volume 100, Pages 235–246, ISSN 0959-6526, https://doi.org/10.1016/j.jclepro.2015.03.055.

25. T. Wagner, C. Herrmann, S. Thiede (2017) Industry 4.0 impacts on lean production systems. Procedia CIRP 63(2):125–131.

Chapter 4

Machine learning assisted manufacturing

*Arvind K. Gupta, Arun Kumar,
and Naman K. Pande*

4.1 INTRODUCTION

The process of goods fabrication from raw materials by the means of manpower, equipment, machines, etc. is known as manufacturing [9]. Manufacturing can range from small hand-crafted items like bangles, confectionery, and hand looms, etc. to extremely high technology-oriented items like semiconductors, integrated circuits, and satellites, etc. The process of manufacturing initiates with the selection of design and raw materials for the target product. Production of this target product then goes through a series of intermediary procedures until the desired product is generated which meets satisfactory quality standards [15, fig. 4.1]. Manufacturing has plenty of implications for the modern world not only in satisfaction of the demand and supply but also as the key driving force of the world economy. Therefore, the process of manufacturing needs to be smooth and problem free and hence needs to evolve with technological advancements.

With the fourth industrial revolution, the rise of "Industry 4.0" took place. Industry 4.0 allowed industries to become "smart" which is crucial for modern industries to survive in the ever-competition-hungry world. Industry 4.0 brought about an evolution in the organisation and control over various manufacturing processes thus leading to an enhanced quality of goods with much-improved efficiency and life cycle. Industry 4.0 includes the incorporation of the latest technology into the process of manufacturing by allowing the use of smart sensors that aid in the collection of meaningful data, the internet of things (IoT), cloud-based manufacturing, and robots among other things. With the acceptance of technology in modern manufacturing, the need for automation of various tasks brought about the need for incorporating artificial intelligence in manufacturing processes.

The term machine learning was coined in 1959, by an American scientist, Arthur Samuel, the pioneer of the fields of artificial intelligence and computer gaming. Machine learning, broadly speaking, is a process that allows any particular system to learn the behaviour of some phenomenon with the help of known data. From this data, the learning algorithms try

DOI: 10.1201/9781003246466-4

77

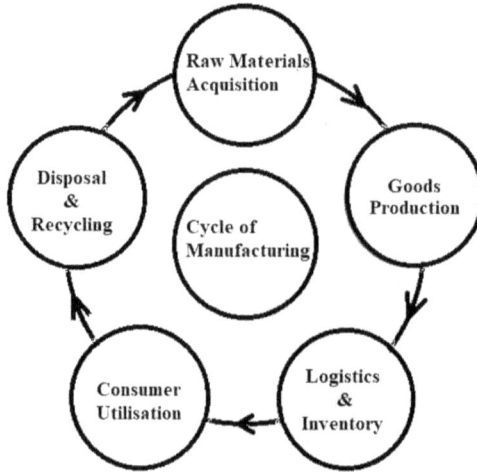

Figure 4.1 Shows the fundamental steps involved in a manufacturing process

to figure out meaningful patterns and knowledge. It uses various mathematical models to figure out the trends in the data. In the current age of information, there is an unprecedented influx of data whose advantage is effectively taken up by machine learning algorithms to carry out the tasks of prediction, segmentation, recommendations, and anomaly detection among others. The learning algorithms are designed in such a way that they can carry out all the above tasks without explicitly being programmed to do so. Machine learning was developed in the quest of developing artificial intelligence. In the 1990s machine learning was later revamped and its goal became to handle real-world problems sequentially.

Apart from statistics and probability, one of the most important foundations of machine learning is optimisation. Most of the learning algorithms are designed in such a way that they have to minimise some error or loss function that is defined for a given set of training data. With the fast-paced advancement in technology, the growth in the availability of data has been tremendous. Hidden in this data is invaluable knowledge, like patterns, signals, and other important messages, that, if figured out, can help in improving the human understanding of how our world works. The field of machine learning had a very powerful impact on the development of the manufacturing industry. With the industrial revolution and the current paradigm of artificial intelligence, backed by the so-called Industry 4.0, which encourages the usage of smart sensors, devices, and machines, the process of data collection has become quite easy and efficient and has led to the increase in efficiency of the manufacturing process. The onset of machine learning algorithms has helped in generating valuable insights which allow the discovery of various complex patterns and allows manufacturers to make intelligent decisions with mathematically backed logic.

4.2 INDUSTRY 4.0

Industry 4.0 is based on the premise that there will always be huge availability of datasets to gain information from. The existence of large and complex datasets, called "big data," is of utmost importance for the proper functioning of industry 4.0 [10]. Big data can be processed only by the means of advanced technologies, algorithms, or software as the conventional techniques fail miserably to gain insights from this huge amount of data. Big data has greatly altered the manufacturing processes by incorporating various interdisciplinary areas like computer science, mathematics and advanced statistics and so on (see e.g. [23–25] and references therein).

Industry 4.0 was kickstarted by the fast-paced advancements in technology like information and communications technology (ICT) and IoT. Applying both the cyber-physical systems (CPS) and the IoT has led to the development of intelligent, flexible systems that can self-learn from the environment they are exposed to. That is the core of Industry 4.0. These goals can only be realised by processing the big data and to achieve this there arise the need to implement machine learning by the means of data mining, statistics, rule mining, pattern recognition, and various other methods [3]. Due to these reasons, machine learning has become an intrinsic characteristic of Industry 4.0 and is implemented in almost every major aspect of the manufacturing processes, to derive as much information from the data as possible in a very short amount of time. This helps in generating meaningful decisions for the companies so that they can maximise their gains [15].

4.3 MACHINE LEARNING

In this section, some important machine learning algorithms are discussed briefly which are ubiquitous in the manufacturing industry.

4.3.1 What is machine learning?

> Any process by which a system improves its performance is "Machine Learning"
>
> *-Herbert Simon (1970)*

Machine Learning is the study of various algorithms which depend on the data available to improve their performance [1]. ML algorithms mainly comprise formulating a mathematical model with the help of sample data, referred to as the training data, and applying various optimisation techniques to devise an efficient predictive model.

The main task of this predictive model is to estimate the outcome with the least possible error, which is tested on a different sample of data called the testing data.

Based on the type of data, the field of machine learning has the following paradigms(see Figure4.2):

- **Supervised learning:** This is the set of learning algorithms which work on the data which contains both the inputs and outputs, also called the labelled data [2]. In these methods, the dataset is a tuple of form $\{(x_i, y_i)\}$, $i = 1, 2, ..., N$, where x_i is called the input and y_i is called the output corresponding to the input x_i. These algorithms try to learn a function $y = f(x)$ such that for each $i = 1, 2, ..., N$

$$y_i = f(x_i) + \in_i,\qquad\qquad(4.3.1)$$

- where $\in_i$ denotes the tolerable error in the prediction of the output, y_i, of the input, x_i.
- Depending on the output y_i, these algorithms can further be classified into 2 categories
    - **Classification:** If each of the y_i belongs to a set of finite and meaningful symbols. Example: {Yes, No}, {Male, Female}, recognising images of animals, therefore the output can be {Lion, Tiger, Cheetah, Leopard} etc.
    - **Regression:** If each y_i belongs to a set of real numbers.
- Example: stock price of NIFTY50 index, blood glucose levels prediction of heart patients etc.
- **Unsupervised learning:** These algorithms work on the data which contains only the inputs and no output y is present, also called the unlabelled data. These algorithms try to partition the given data

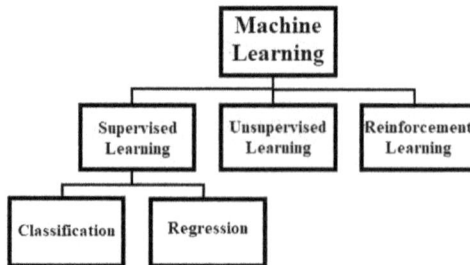

Figure 4.2 Graphical illustration of the machine learning paradigms

$\{x_i\}_{i=1}^{N}$ into disjoint groups called clusters, such that, the points in a single cluster are as similar as possible, while points in two distinct clusters are as dissimilar as possible. Examples: Market research and customer segmentation for a services-based company.

- **Reinforcement learning:** This entails defining a system to generate a sequence of decisions required to carry out some specific tasks. The reinforcement learning algorithms try to simulate a game-like situation, in which various trial and error methods are employed to figure out a solution. The algorithms based on reinforcement learning techniques also used unlabelled data. Examples: Self-driving cars, chess games simulation, etc.

The architecture of the machine learning models has three components.

- **Representation:** Characterisation of what is being learnt. That is, formulating a mathematical model using the sample data.
- **Evaluation:** Measure the goodness of what is being learnt. That is, to formulate an error function which serves to evaluate the predictive capability of the model being deployed.
- **Optimisation:** Given the representation and evaluation metric, figure out the optimum representation of the proposed model.

4.3.2 Support vector machines

It is one of the most popular supervised learning algorithms used for both regression as well as classification purposes. It is most popular for the latter but proves to be quite useful for regression tasks as well. For classification purposes, its main idea is to figure out a decision boundary between the different classes of data [1].

The main objective of the SVM is to find an optimal hyperplane that maximises the gap or margin between the classes.

Definition (hyperplane): A hyperplane in d dimensions is the set of points $x = (x_1, x_2, x_3, ..., x_d) \in R^d$ that satisfy the equation,

$$\beta_1 x_1 + \beta_2 x_2 + \beta_3 x_3 + ... + \beta_d x_d + c = 0 \qquad (4.3.2)$$

where, $\beta_1, \beta_2, ..., \beta_d$ are called the weights and c is called the bias.

For example, in R^2 a hyperplane is nothing but a line, while in R^3, a hyperplane is a plane (Figure 4.3).

Definition (halfspace): Consider the hyperplane in d

$$P(x) = \beta_1 x_1 + \beta_2 x_2 + \beta_3 x_3 + ... + \beta_d x_d + c = 0$$

Then this hyperplane $H(x)$ divides the plane into two halfspaces defined as

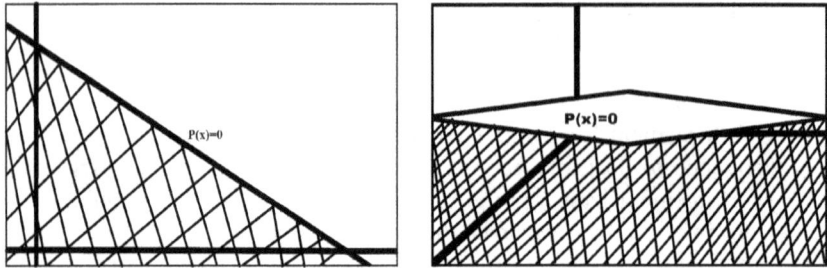

Figure 4.3 Illustrates the hyperplanes in R^2 and R^3 and the distinct half spaces generated by them

$$P_+(x) = \beta_1 x_1 + \beta_2 x_2 + \beta_3 x_3 + \ldots + \beta_d x_d + c \geq 0 \qquad (4.3.3)$$

called the upper halfspace(Figure4.3) and

$$P_-(x) = \beta_1 x_1 + \beta_2 x_2 + \beta_3 x_3 + \ldots + \beta_d x_d + c < 0 \qquad (4.3.4)$$

Called the lower halfspace(Figure4.3).

4.3.2.1 Maximal margin classifier

With the assumption that the given labelled data is linearly separable, the task of the maximal margin classifier is to find a hyperplane that separates the data points into two distinct half spaces generated by this hyperplane.

Let $\beta = (\beta_1, \beta_2, \ldots, \beta_d)^T$ be the weights vector and c be the bias, then for any $X = (X_1, X_2, \ldots, X_d) \in R^d$, let

$$P(X) = \beta_1 X_1 + \beta_2 X_2 + \ldots + \beta_d X_d + c = \beta^T X + c = 0$$

For a given dataset $D = \{(x_i, y_i)\}_{i=1}^{N}, x_i \in R^d$ and $y_i \in \{-1, 1\}$, we want to figure out β and c such that the following two conditions hold

$$\beta^T x_i + c \geq 1, if\ y_i = +1 \quad \beta^T x_i + c \leq 1, if\ y_i = -1 \qquad (4.3.5)$$

$$\Rightarrow y_i (\beta^T x_i + c) \geq 1 \forall\ i = 1, 2, 3, \ldots, N \qquad (4.3.6)$$

Now for the hyperplane $P(X) = 0$, we have the two parallel hyperplanes

$$P_1(X) = \beta^T X + c = +1 \qquad (4.3.7)$$

and

$$P_2(X) = \beta^T X + c = -1 \qquad (4.3.8)$$

Definition (margin): The distance of these two parallel hyperplanes $P_1(X) = +1$ and $P_2(X) = -1$ from the optimal separating hyperplane $P(X) = 0$ is called the **margin** of the maximal margin classifier [**fig. 4.4.**].

Definition (support vector): Any datapoint x_i that lies on any of the parallel hyperplanes $P_1(X) = +1$ or $P_2(X) = -1$ is called the **support vector**(Figure4.4).

With the help of these support vectors, the problem of the maximal margin classifier comes down to maximising the margin or equivalently the distance between the two hyperplanes, $P_1(X) = +1$ and $P_2(X) = -1$.

Now, the distance between the two hyperplanes $P_1(X) = \beta^T X + c = +1$ and

$P_2(X) = \beta^T X + c = -1$ is given by,

$$\delta = \frac{1}{\| \beta \|} - \frac{-1}{\| \beta \|} = \frac{2}{\| \beta \|} \qquad (4.3.9)$$

where $\| \beta \|$ is the **Euclidean norm** defined as $(\beta^T \beta)^{\frac{1}{2}} = \left(\Sigma_{i=1}^{d} \beta_i^2\right)^{\frac{1}{2}}$

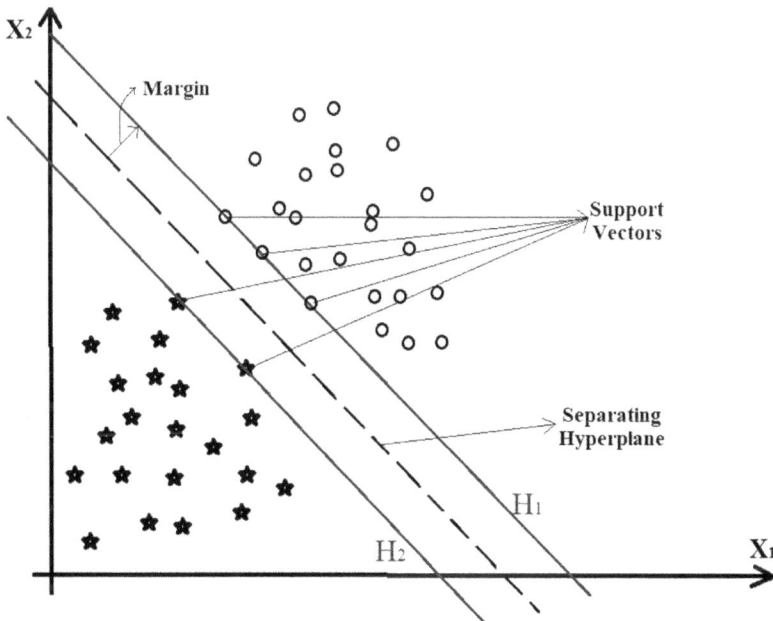

Figure 4.4 Illustrates an example of a maximal margin classifier

Therefore, the problem becomes

$$\frac{2}{\|\beta\|} \ s.t. \ y_i\left(\beta^T x_i\right) \geq 1 \forall i = 1, 2, ..., N \tag{4.3.10}$$

Now maximising $\frac{2}{\|\beta\|}$ is the same as minimising $\frac{\|\beta\|}{2}$, which is equivalent to minimising $\frac{\|\beta\|^2}{2}$, therefore, we get an equivalent optimisation problem,

$$\|\beta\|^2 \ s.t. \ y_i\left(\beta^T x_i\right) \geq 1 \forall i = 1, 2, ..., N \tag{4.3.11}$$

which is preferred as our problem has converted into a convex optimisation problem.

Now define the Lagrangian of the above optimisation problem as

$$L = \frac{\|\beta\|^2}{2} - \sum_{i=1}^{n} \lambda_i(y_i\left(\beta^T x_i + c\right) - 1 \tag{4.3.12}$$

where $\lambda_i, i = 1, 2, ..., N$ are called Lagrangian multipliers. By taking the partial derivative of L, w.r.t β and c and equating them to zero we get the following Karush Kuhn Tucker (KKT) conditions.

$$\beta = \sum_{i=1}^{N} \lambda_i y_i x_i, \sum_{i=1}^{N} \lambda_i y_i = 0, \lambda_i\left(y_i\left(\beta^T x_i + c\right) - 1\right) = 0 \forall$$

$$i = 1, 2..., N, \lambda_i \geq 0 \forall i = 1, 2, 3, ..., N \tag{4.3.13}$$

With the help of the above equations, we can formulate the dual of the above problem as

$$\sum_{i=1}^{N} \lambda_i - \frac{1}{2} \sum_{i=1}^{N} \sum_{j=1}^{N} \lambda_i \lambda_j y_i y_j x_i^T x_j \ s.t. \lambda_i \geq 0, \forall i = 1, 2, ..., N \sum_{i=1}^{N} \lambda_i y_i = 0 \tag{4.3.14}$$

where $\lambda = (\lambda_1, \lambda_2, ..., \lambda_N)$.

This dual formulation allows us to evaluate the Lagrangian multipliers λ_i.

Now, by KKT conditions

$$\lambda_i\left(y_i\left(\beta^T x_i + c\right) - 1\right) = 0 \forall i = 1, 2, ..., N,$$

we have two cases:

$\lambda_i = 0$ or

$$\lambda_i \geq 0 \Rightarrow y_i\left(\beta^T x_i + c\right) = 1$$

This implies that if $\lambda_i \geq 0$, then x_i must be a support vector and if $\lambda_i = 0$, then, $y_i\left(\beta^T x_i + c\right) > 1$ implying that x_i is not a sssupport vector.

Therefore, from this observation, we get that

$$\beta = \sum_{\lambda_i > 0} \lambda_i y_i x_i \tag{4.3.15}$$

This implies that the weight vectors are nothing but a linear combination of the support vectors.

Now to compute the bias we calculate c_i per support vector x_i as follows,

$$\lambda_i\left(y_i\left(\beta^T x_i + c_i\right) - 1\right) = 0 \tag{4.3.16}$$

since, $\lambda_i > 0$ for each support vector x_i we get that,

$$y_i\left(\beta^T x_i + c_i\right) - 1 = 0 \tag{4.3.17}$$

$$\Rightarrow c_i = \frac{1}{y_i} - \beta^T x_i = \frac{y_i^2}{y_i} - \beta^T x_i = y_i - \beta^T x_i \left(y_i^2 = 1 \, as \, y_i \in \{-1,1\}\right) \tag{4.3.18}$$

$$\Rightarrow c_i = y_i - \beta^T x_i \tag{4.3.19}$$

Therefore, we can take c as the average bias of all c_i's for each support vector x_i.

$$c = average_{\lambda_i > 0}\left\{c_i\right\} \tag{4.3.20}$$

Example 1: Consider the dataset
Formulate the optimisation problem for the data in **table 4.1**[see **fig. 4.5**] as

$$\frac{\|\beta^2\|}{2} \quad s.t. y_i(\beta^T x_i + c) \geq 1 \forall i = 1,2,\ldots,9 \tag{4.3.22}$$

The Lagrangian of the above problem becomes

Table 4.1 Given table represents the classification of two-dimensional data into classes I and -I

Index	X_1	X_2	Y
x_1'	2	4	I
x_2'	2	2	I
x_3'	4	4	I
x_4'	I	3	I
x_5'	2	3	I
x_6'	2	I	-I
x_7'	4	3	-I
x_8'	3	I	-I
x_9'	4	2	-I

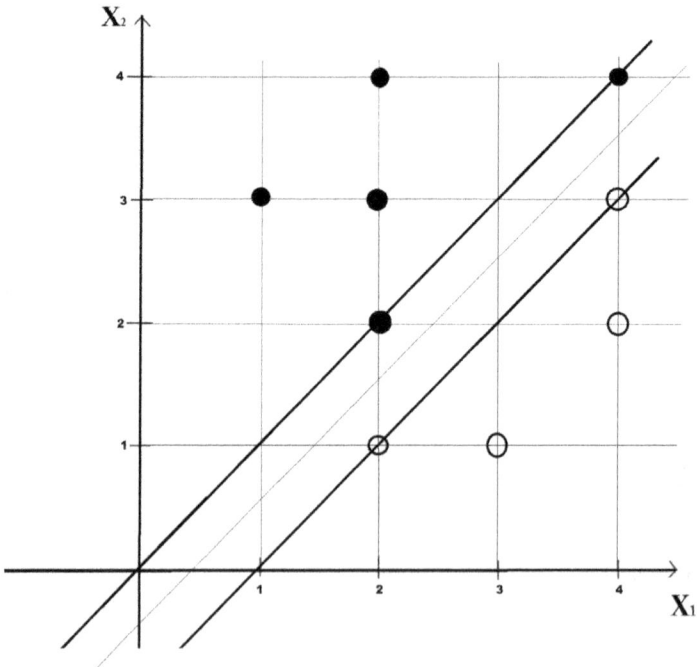

Figure 4.5 Illustrates the data given in Table 4.1 graphically

$$L = \frac{\|\beta\|^2}{2} - \sum_{i=1}^{7} \left(\lambda_i \left(y_i \left(\beta^T x_i + c \right) - 1 \right) \right) \tag{4.3.33}$$

From the condition that $\sum_{i=1}^{7} \lambda_i y_i = 0$, we get

$$\lambda_1 + \lambda_2 + \lambda_3 + \lambda_4 + \lambda_5 - \lambda_6 - \lambda_7 - \lambda_8 - \lambda_9 = 0 \tag{4.3.34}$$

Further, we have that

$$\lambda_i \left(y_i \left(\beta^T x_i + c \right) - 1 \right) = 0 \, \forall i = 1, 2, \ldots, 9 \tag{4.3.35}$$

To figure out the support vectors we would apply the graphical approach as solving the above set of equations is computationally expensive and requires the use of the software.

From the figure, it is evident that the support vectors are $(2,2)$, $(4,4)$, $(2,1)$, $(4,3)$

Therefore, we get that

$$\lambda_1, \lambda_4, \lambda_5, \lambda_8, \lambda_9 = 0 \, \lambda_2, \lambda_3, \lambda_6, \lambda_7 > 0 \tag{4.3.36}$$

Hence, we have

$$2\beta_1 + 2\beta_2 + c - 1 = 0 \, 4\beta_1 + 4\beta_2 + c - 1 = 0 \, 2\beta_1 +$$
$$\beta_2 + c - 1 = 0 \, 4\beta_1 + 3\beta_2 + c - 1 = 0 \tag{4.3.37}$$

Solving the above equations yield

$$\beta_1 = -2, \ \beta_2 = 2 \ and \ c = 1 \tag{4.3.38}$$

Therefore, the equation of the maximal margin classifier becomes

$$P(x_1, x_2) = -2x_1 + 2x_2 + 1 = 0 \tag{4.3.39}$$

Therefore, any point for which $P(x_1, x_2) < 0$ will be classified as -1, and any point for which $P(x_1, x_2) > 0$ will be classified as 1.

4.3.2.2 Support vector machine classifier

One of the inherent problems with the maximal margin classifier is the assumption that the data will be linearly separable, which is not always the case when dealing with real-world data [26]. Another problem with the maximal margin classifier is the heavy reliance on the support vectors. Therefore, if the support vectors change then the classifier changes significantly and hence may lead to overfitting. Due to these reasons, the maximal margin classifier is also referred to as a hard margin classifier.

The support vector machine or soft margin classifier was developed as a solution to the underlying problems of the maximal margin classifier, which allows some misclassification in the classification.

In the maximal margin classifier, the separating hyperplane can be obtained if we set the condition $y_i(\beta^T x_i + c) \geq 1$. The way SVMs deal with the nonlinearity in the data with the introduction of a slack variable s_i, i.e.,

$$y_i(\beta^T x_i + c) \geq 1 - s_i \qquad (4.3.40)$$

where $s_i \geq 0$ is the slack variable for point x_i. The slack variable s_i represents the amount by which point x_i violates the separability condition. If $s_i = 0$, then this implies that the point x_i is at least $\dfrac{1}{\|\beta\|}$ distance away from the separating hyperplane. If $s_i > 0$, then the point is not at least $\dfrac{1}{\|\beta\|}$ distance away from the hyperplane. If $0 < s_i < 1$, then the point x_i is still on the correct side of the hyperplane. If $s_i > 1$ then point x_i is completely misclassified and lies completely on the wrong side of the hyperplane.

The new objective function of the SVM becomes,

$$\frac{\|\beta\|^2}{2} + S\sum_{i=1}^{N}(s_i)^k \ \text{s.t.} \ y_i(\beta^T x_i + c) \geq 1 - s_i, \forall i = 1, 2, ..., N$$
$$s_i \geq 0 \forall i = 1, 2, ..., N \qquad (4.3.41)$$

where S and k are constants. The term $\sum_{i=1}^{N}(s_i)^k$ gives the loss.

The constant S is called the regularisation constant, which is used to control the maximisation of the margin and minimisation of the error. The smaller the value of S, the objective function will be aimed towards maximisation of the margin, the larger the value of S, the margin would have less control over the objective and the aim of the objective would be to minimise the loss. The constant k defines the type of loss function used. For $k = 1$, the loss is called the hinge loss and if $k = 2$, the loss is called the quadratic loss.

The Lagrangian function for the soft margin classifier is defined as

$$L = \frac{\|\beta\|^2}{2} + S\sum_{i=1}^{N}(s_i)^k - \sum_{i=1}^{N}(\lambda_i(y_i(\beta^T x_i + c) - 1 + s_i) - \sum_{i=1}^{N}\mu_i s_i \quad (4.3.42)$$

where λ_i and μ_i, $i = 1, 2, ..., N$ are the Lagrangian multipliers.

Similar to solving the optimisation problem for the maximal margin classifier, the weights of the hyperplane for both the hinge and quadratic losses are given by

$$\beta = \sum_{\lambda_i > 0} \lambda_i y_i x_i \tag{4.3.43}$$

and the bias term is given by

$$c = average\{y_i - w\beta^T x_i\} \tag{4.3.44}$$

4.3.3 Artificial neural networks

Artificial neural networks are among the most effective learning methods currently known. It provides an excellent solution for approximating the real of discrete or vector-valued target functions with a high level of accuracy. ANN model proves to be quite robust for problems, such as interpreting highly complicated data provided by sensors [6].

Artificial neural networks prove to be a great model for supervised learning for classification as well as regression. Neural networks are useful for approximating any function with a very high accuracy, which is capable of interpreting extremely complex data.

The training data that correspond to inputs from sensors such as cameras and microphones are quite noisy, and complex. This complexity and noise are not easy to handle. ANN proves to be extremely well suited to deal with such problems.

The most suitable applications for ANN learning are to interpret meaningful knowledge from complex, noisy data that is obtained from sensors like cameras, and microphones [6].

4.3.3.1 Feedforward network and backpropagation algorithm

In general, the architecture of an artificial neural network is composed of three layers:

A. Input layer.
B. Hidden layer.
C. Output layer.

The neural network can be considered a directed graph $G = (V, E)$ where V is the set of vertices called the neurons and E is the set of edges that connects these neurons by some weights [see fig. 4.6]. The first layer or the

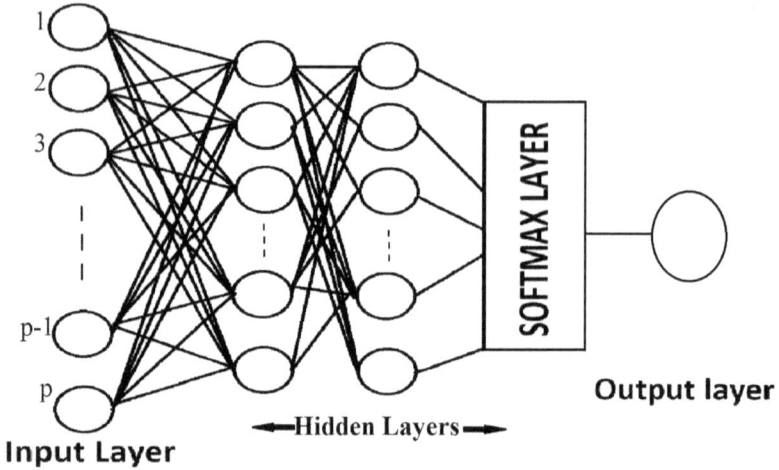

Figure 4.6 Illustrates the general architecture of a feedforward neural network with hidden layers and a softmax activation function used for classification

input layer is composed on N neurons whose values are nothing but data samples in the data set. Each neuron in the input layer takes the data in the form of a vector $x \in R^d$. This input is passed onto a series of, k, hidden layers with the help of some linear or non-linear transformations.

$$H_l = \sigma\left(\beta_l H_{l-1} + c_l\right), \forall l = 1, 2, \ldots, D \tag{4.3.45}$$

where $H_0 = X$, and $\beta_l \in R^{nH_l \times nH_{l-1}}$ is the weights for the hidden layer l and c_l is the bias and σ is either a linear or non-linear activation function and n_{H_j} = number of nodes in hidden layer $j, j = 1, 2, \ldots, D$.

After calculating the output of the last hidden layer, k, the output is passed through a $(d+1)^{th}$ layer called a softmax layer (for classification problems), i.e.,

$$H_{D+1} = \beta_{D+1}^T H_D + c_{D+1} \tag{4.3.46}$$

and a non-linear activation function is applied to the output of this layer. This non-linear function calculates the probability of each neuron using the softmax function given by,

$$\hat{H} = \frac{e^{H_{(D+1)j}}}{\sum_{i=1}^{p} e^{H_{(D+1)i}}} \tag{4.3.47}$$

where $p = n_{d+1}$ is the number of neurons, $H_{(D+1)k}$ is the output value of the kth neuron in the $(D+1)^{th}$ layer.

After applying the softmax function the output is converted to a probability distribution, based on which the class label for the input x is assigned the class that has the maximum probability. The feedforward network sets up the basic architecture of neural networks. But the main task for setting up the model for real-life applications, the parameters of this network, $\beta_l, c_l \ l = 1, 2, ..., D$ must be learnt. To handle this task, we apply the backpropagation algorithm to our feedforward network which learns these parameters with the help of the gradient descent algorithm. Concerning a loss function, the gradient of the network is computed in the weight space. The main aim of the backpropagation algorithms is to minimise this loss function by trying to obtain the optimal parameters $\beta_l, c_l \forall l = 1, 2, ..., D+1$.

Example 2: Construct an artificial neural network that models the XOR gate.

Consider the data table

The XOR gate requires two hidden layers along with an input layer to be modelled [see fig. 4.7].

The neural network weights for the first hidden layer units are $\beta_{11}^{(1)} = 20$, $\beta_{12}^{(1)} = -20$, $\beta_{21}^{(1)} = 1$, $\beta_{22}^{(1)} = -20$ and $c_1^{(1)} = -10, c_2^{(1)} = 30$.

For the second hidden layer, the weights are $\beta_{11}^{(2)} = 20$, $\beta_{21}^{(2)} = 20$ and $c_1^{(2)} = -30$.

The sigmoid activation function is used for each computation,

$$\sigma(x) = \frac{1}{1 + e^{-x}} \tag{4.3.48}$$

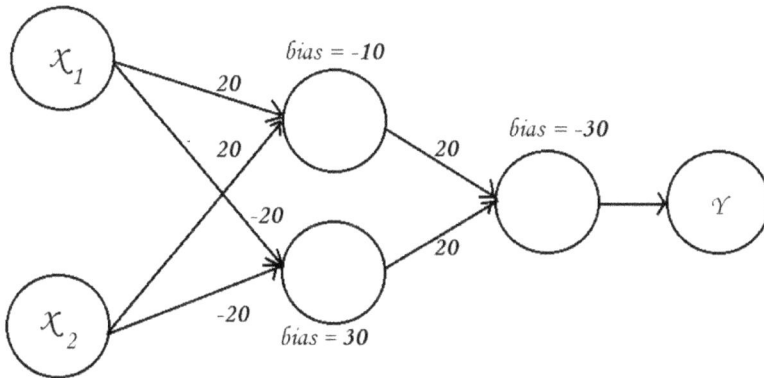

Figure 4.7 Graphically represents a two hidden layer network that models the XOR gate for the data provided in the Table 4.2

Table 4.2 The table shows the logic gate XOR

X_1	X_2	Y
0	0	0
0	1	1
1	0	1
1	1	0

For $(X_1, X_2) = (0,0)$ we have the following computations for the hidden layer 1:

For the first node in first hidden layer, we get

$$\sigma\left(\beta_{11}^{(1)} \times 0 + \beta_{21}^{(1)} \times 0 + c_1^{(1)}\right) \sim 0, \qquad (4.3.49)$$

Therefore, the hidden node takes the value 0.

For the second node in the first hidden layer, we get

$$\sigma\left(\beta_{12}^{(1)} \times 0 + \beta_{22}^{(1)} \times 0 + c_2^{(1)}\right) \sim 1 \qquad (4.3.50)$$

Therefore, the hidden node takes the value 1.

These two values are passed onto the next hidden layer. The computations for the next hidden layer are as follows:

For $\left(H_1^{(1)}, H_2^{(1)}\right) = (0,1)$ we get

$$\sigma\left(\beta_{11}^{(2)} \times 0 + \beta_{21}^{(2)} \times 1 + c_1^{(2)}\right) \sim 0 \qquad (4.3.51)$$

Therefore, we get $0 \oplus 0 = 0$. Similarly, the other computations are done.

4.3.3.2 *The forward and the backward pass*

The feedforward network of neurons with a d dimensional input layer, d D+1 hidden layers, and a softmax layer, as an output layer are defined with the help of interconnected neurons with a set of weights and thresholds which need to be figured by the backpropagation algorithm.

4.3.3.2.1 *Forward pass*

The computation rule for the forward pass can be summarised as follows: Set $H_0 = x$. For layers $l = 1, 2, \ldots, D+1$ recursively set,

$$H_l = \sigma_l \left(\beta_l H_{l-1} + c_l \right), \; l = 1,2,3,\ldots,D \; \hat{H} = F\left(H_{D+1} \right) \tag{4.3.52}$$

where $\hat{H}$ is the predicted output for the input vector x, σ_l is the activation function for the lth layer and F is the softmax activation function.

After setting up the feedforward network we apply the backpropagation algorithm with the help of the gradient descent algorithm.

The learning rule for the backward propagation algorithm can be summarised as follows:

4.3.3.2.2 Backward pass

Define the loss function as L and set

$$H_0 = x \, , \hat{H} = H_{D+1} \tag{4.3.53}$$

1. For the $(D+1)^{th}$ layer compute $\nabla_{H_{D+1}} L = \nabla_{\hat{H}} L$
2. For layers $D, D-1,\ldots,1$
   a. Compute

   $$J_{H_l} \left(\sigma^{(-1)} (H_l) \right) \tag{4.3.54}$$

   the Jacobian for the $(l)^{th}$ layer
   b. Recursively compute

   $$\sigma^{-1}(H_l) L = L(J)_{H_l} \left(\sigma^{-1}(H_l) \nabla_{H_{l-1}} (L) = \nabla_{\sigma^{(-1)(H_l)}} L(\beta_l) \right) \tag{4.3.55}$$

   c. Computing gradients

   $$\nabla_{\beta_l} (L) = H_{l-1} \nabla_{\sigma^{-1}(H_l)} (L) \nabla_{c_l} (L) = \nabla_{\sigma^{-1}(H_l)} (L) \tag{4.3.56}$$

Update the weights by the learning rule

$$\beta_l = \beta_l - \eta \nabla_{\beta_l} (L) c_l = c_l - \eta \nabla_{c_l} (L) \tag{4.3.57}$$

Recursively keep on applying the forward and the backward pass with the weight updates until desired accuracy or the number of iterations is achieved.

The artificial neural networks model considers the factors, such as time in building the model, the training data, and global as well as the partial characteristics of the data that may influence the errors in prediction. There are a lot of uses for applying artificial neural networks to a complex manufacturing process speed of processing huge amounts of information, evolving with the type of problem at hand and gaining insights as required, ANN are quite robust, and can work with problems where space and power restrictions are present.

4.4 MACHINE LEARNING IN MANUFACTURING PROCESSES

The processes of manufacturing can be an extremely daunting task for industries, especially the ones that have just risen up and do not have enough capital to carry out the manufacturing process smoothly. If not carried out properly due to lack of expertise or lack of adequate machinery, the quality of the manufactured goods can become inadequate thus leading to low sales and hence profits and may also adversely affect the reputation of the company. With the help of machine learning, the process of manufacturing has become "smart" and mundane tasks have been automated that has minimised errors to a very large extent [11,12]. With the application of machine learning in the manufacturing process, we can observe a huge rise in the productivity of the industries, helping in both business and employment generation opportunities. Machine learning can aid the development of novel and fantastic business strategies with minimal human interference for these industries.

Artificial intelligence has facilitated the manufacturing industries in the following ways.

4.4.1 Machinery malfunction

Even though the arduous tasks required in manufacturing processes are carried out quite effectively by machines, it does not imply that the processes would always run smoothly and efficiently. The processes involved in manufacturing cause wear and tear in these machines. Although the breakdown or malfunction of these machines is inevitable, there are still certain measures that can be taken up to avoid these malfunctions. There are various maintenance techniques that industries apply to operate smoothly. Most industries around the globe deploy **corrective maintenance measures,** in which the damaged, worn-out parts are replaced as soon as they don't provide useful functionality to the machines. Although this strategy is quite straightforward and hassle free, it is by far the most expensive as the procurement of new parts is not quite easy. **Preventive maintenance** measures are quite robust than corrective measures. The measures, just like the corrective ones, replace the damaged parts albeit, these measures don't wait for the parts to become completely useless. This method allows the minimisation of unwarranted maintenance costs and also avoids devastating breakdowns. **Predictive maintenance** tries to create a balance between both the corrective and preventive maintenance measures with the help of automating the process of invigilation of machines.

Automating the maintenance of machines, and predicting when the correct point of time when the machines can malfunction, has allowed the manufacturers to devise and implement appropriate strategies to deploy machines maintenance plans on time, thus allowing the machines to work smoothly and effectively.

4.4.2 Product quality management

Inspection of the quality of finished products is one of the most important tasks for any manufacturing industry as the reputation and the profits of the company depend upon how good the finished products of the company are and how well the customers perceive those products. It is one of the processes of manufacturing that makes certain that the quality of the products that are manufactured is either consistent or can be further improved. This process needs to maintain the industry's desire for perfection because any lax in quality management can lead to deterioration of the quality of the finished product, the profits and most importantly, the reputation of the company that relies on how the products are perceived. For example, the food and pharmaceuticals industries are heavily reliant on the quality of the finished products, because if a defective item is rolled out to the public, the results can be devastating for the consumer as well as the manufacturer as both the parties, in this case, is sensitive to the quality of the finished product. Therefore, the need for skilled personnel who strive for perfection is the most important criterion for ensuring above-par efficiency.

Quality control established a set of controls that standardise the manufacture of the goods and the general notion of the item among the consumers. Traditionally this was achieved by deploying various personnel at different stages of the manufacture of goods. This minimises the room for errors as only those personnel are selected who have adequate training can ensure an error-free manufacturing process. Although this ensures that the quality of the goods would be satisfactory, the deployment of so many personnel is not always possible, especially for industries with low capital. Further no matter the amount of training received, the amount of time taken to process the huge amount of information and data can hamper quality control measures. These reasons, among others, led to the introduction of artificial intelligence and machine learning in the quality control process. With the deployment of artificial intelligence at various stages of quality checks, huge amounts of data can be processed within seconds and help in filtering out the flawed products from the flawless ones, thus, eliminating the human errors that can arise from the lack of expertise or mundaneness of the task.

4.4.3 Logistics and inventory

Logistics is the process of management of goods, raw or finished, acquisition, storage, and transportation of resources to the final destination to meet the demands. Logistics management entails figuring out the potential suppliers and the sellers along with the effectiveness and ease of access to the market. Its goal is to effectively and timely figure out the optimum amount of resources at the correct time and transportation of the said resources to the final destination in an orderly and timely fashion. For example, for a natural gas company, the goal of the logistics management is to maintain the gas pipelines, the transportation truck that moves around the oil that is transformed into the natural gas along the supply chain, and management of the distribution centres. This example underlines the significance of logistics in the manufacturing process. If the logistics management is not carried out perfectly, it can lead to untimely deliveries, and failure to meet the demands of the consumers which ultimately leads to loss of capital and hamper the reputation of the companies, no matter how well the quality control process was carried out.

Another important consideration for the manufacturing industries along with logistics management is inventory management. The inventory of a company is defined as the complete list of items that the company possesses, like property, goods in stock, etc. Inventory management is the process of storing, using and selling a company's inventory. It manages the raw materials required for the manufacture of the goods, the components required for maintaining the machines, and the finished products that were manufactured and warehouses for the storage of these items. The inventory of a company can be considered both an asset as well as a liability. For example, for the manufacturing industries, the raw materials and the finished products are the heart and soul and cannot operate without these two items. Therefore, if not managed properly it will hamper the manufacturing processes. On the other hand, a large inventory is always at risk of being stolen, or damaged due to unprecedented circumstances and the finished products are prone to shifts in demands of the market, for example, the fashion industry, further, all the inventory must always be insured to minimise the losses borne by the companies.

The manufacturing industry needs to take utmost care of logistics and inventory management as these are two of the most important tasks in manufacturing because errors in either logistics or inventory can lead to a huge loss to the company. Automating these two tasks can help in boosting the efficiency of the logistics and inventory, thus increasing the revenue of the industries. Applying machine learning in logistics and inventory management can yield positive results for the industries. It can aid in cutting costs by systematically getting rid of waste and improving the quality of the storage units. It can optimise the amount of inventory that should be maintained thus eliminating the need to store huge stockpiles of resources and finished products. One of the most problematic things that plague industries in holding inventories is the fraudulent behaviour of the personnel, by applying machine learning the detection of fraud can be done quite easily.

4.4.4 Cybersecurity

Cybersecurity is the field of safeguarding highly sensitive systems, networks, etc. from phishing, malware, and ransomware cyberattacks. The target of these cyberattacks is usually to extract huge ransoms, gain access to or destroy highly sensitive and protected data, or just hamper the normal business cycles. Therefore, it is extremely necessary to systematically implement cybersecurity measures so as to prevent the destruction of data or to prevent data from leaking into the hands of unwanted users. Common cybersecurity threats include malware which refers to "malicious software" variants like viruses, trojans, and spyware. The malware tries to gain unauthorised access to private networks or computer systems by the means of uncertified software or websites. Another threat is ransomware which is malware that tries to sabotage the computer systems or networks by locking down the files/folders that contain highly sensitive data that the hacker and/or the user deemed important. These files are unlocked only if a certain amount of ransom is paid. Phishing tricks the users into leaking their data to the hacker without their information. Phishing scams occur quite naturally and seem quite innocent as they appear in the form of messages or emails from big companies or banks that persuade the users to give their personal information like credit card data, and bank details or may directly extract money from the bank accounts. A good cybersecurity strategy has highly convoluted networks of protections that provide defence against the aforementioned risks. Countermeasures for these attacks may include, critical infrastructure security which includes the protection of computer systems, networks or any other asset upon which the day-to-day processes of citizens relies upon, like, national security, economic conditions, healthcare, and many more. The second domain of cybersecurity may include network security which entails the protection of computer networks from getting attacked via the Internet using both wired and wireless connections. Cloud security ensures that the data stored over the internet in cloud storage remains unhinged to maintain the privacy of the people, communities, or businesses because the leak of the data from this cloud storage can have a detrimental impact on the aforementioned users. For the machine learning algorithms to work effectively, industries need to keep their data and networks safe from possible malware, ransomware, or phishing attacks. Earlier the security and maintenance of these sensitive items could only be done by human experts, but with the advent of robots, AI can allow these robots, with minimal human supervision, to work alongside experts to keep the networks safe and secure. Further many machine learning algorithms as already mentioned are designed to figure out patterns from the data, therefore using the past data of cyberattacks, the algorithms can be implemented to analyse and predict the cyberattacks and either deploy or improve the existing cybersecurity measures available to the company by alarming the relevant authorities. Cybersecurity measures can be improved by making them simpler, proactive, and cost-efficient.

4.4.5 Environment and ecology

Even though the modernisation of industries has impacted the human lifestyle and world economy vastly, this convenience comes at a cost. Industries have caused a detrimental effect on the environment like air pollution, and release of the toxic chemicals in the rivers which causes water contamination. Due to the release of toxic and dangerous materials by the industries into the environment, there is a major setback to the lives of not only humans but also causes problems to the health of various plants and animals and forces them to the verge of extinction. The major problem caused by these industries is air pollution. The manufacturing industries release huge amounts of carbon dioxide and other poisonous gases into the air thus increasing health and environmental damage. This leads to the problem of global warming. As we are already facing shortages in the availability of potable water, water contamination is a major issue that needs to be dealt with, with extreme caution.

Recently there has been a boom in technology that can impact the environment positively. Environmental science machine learning algorithms usually apply supervised learning techniques that help in the prediction of the changes in environmental conditions caused due to manufacturing processes.

Machine learning can aid in the following ways to help the detrimental effects of industries on the environment.

Prediction of particulate matter can be handled extremely effectively by machine learning algorithms when sufficient data is available to learn the relationships between the air pattern and the presence of particulate matter in the air. Prediction of water resource availability. Due to changes occurring in climate, greenhouse gases released into the environment by factories and other sources or changes in rain patterns, a tremendous problem in the prediction of water resources across the regions can rise. Development of decision and support tools that can tackle the problems of variations and uncertainties that are caused due variations in climate or negative interactions between artificial and natural systems are required to manage and sustain water resources. Therefore, to better understand and manage water resources, it has become extremely necessary to employ the 'smart' algorithms that can handle the shifts in natural conditions. Machine learning algorithms can aid in detecting anomalies in the environment that occur due to release of the greenhouse gases or the changes in aquatic life that are caused due to release of toxic chemicals in the stream or in general the changes occurring in health conditions of the animals that are living nearby the factories or be it the labourers working continuously inside the factories [13].

Even though industrial manufacturing has greatly benefited from the applications of AI and ML, the need for a sustainable growth strategy is required which allows the environment and ecology to flourish as well. As most of the power plants around the world are run with the help of fossil fuels, the need for smarter energy consumption is required. With the help of AI, energy-efficient smart plans can be created for industries that need to run 24/7 for optimal profits and efficiency.

4.5 FAULT DIAGNOSTICS USING ARTIFICIAL INTELLIGENCE

One of the most popular and useful applications of AI in manufacturing is the "fault diagnostics in manufacturing." With the availability of various novel machine learning algorithms, we can easily deploy artificial intelligence to figure out the faults in the manufacturing process with minimal human interaction with high precision and accuracy, which ultimately leads to reduction in cost and highly efficient manufacturing process.

There exists a large variety of machine learning algorithms which can aid in diagnosis of these faults in the manufacturing system. This chapter is going to deal with two of the most prominent and highly regarded machine learning algorithms, the support vector machines and the artificial neural networks (ANN), which can aid in the diagnosis of faults in the manufacturing process.

4.5.1 Detecting faults in machinery

Manufacturing cost reduction, machine downtimes reduction and improvement in the quality of the finished product can be well managed with effective equipment monitoring. Some of the most commonly used parts in machines used by the manufacturing industry are the motor pump and ventilator. Rotating elements in the machines tend to have a lot of unavoidable oscillations that are embedded in these components. These oscillations occur with the regular usage of the machines. Due to faults arising in these machines, there is a high chance that some irregularity in the oscillations occurs along with certain other issues. Therefore, it is of utmost importance to pinpoint the location and the existence of these irregularities on time and with high efficiency.

To train the ML algorithms to figure out these faults we need to supply the data to the algorithm which needs to have the following attributes like frequency, velocity, and phase of the harmonic components of the oscillations, etc. The target variable for the problem is machine condition which has the classes like faulty bearings, joint problems, mechanical loosening, distortion in the basement, unbalanced, and normal condition [5].

Assumptions: The above-mentioned problems may arise simultaneously in the real-world situation which makes the task of classification extremely complicated therefore we assume that a single machine can belong to a single type of a class at a time, i.e., only one type of problem may arise in a machine at a time. This assumption is good enough because even if one type of fault is detected in a machine having multiple faults, then measures can be taken automatically to deal with those faults.

4.5.1.1 Applications to thermal power plants

Modern power plants have been computerised, therefore a huge amount of data could be automatically collected and stored. Three major pieces of equipment play a crucial role in the well-ordered functioning of thermal plants. A **steam generator** boils the water to steam and generates optimum pressure and temperature that allows the movement of the electrical generator. The transformation of the thermal energy to mechanical energy is then achieved through the **steam turbine generator**. This mechanical energy is then further converted to electrical energy with the help of an **electrical driven generator**.

Even though the above-mentioned equipment is important, the steam turbine generator is the most important and sensitive equipment in thermal power plants. Therefore, analysing the failures occurring in this equipment is of utmost importance [7].

For predicting failures in steam turbine generators in thermal power plants, the SVM-based prediction model can be used to classify the generators as faulty or working. The training of the SVM model requires a huge amount of data, which may have an extremely large number of attributes, therefore, the model needs to be integrated with feature selection to discard the redundant features which may not have much impact or may prove to be unfavourable for prediction purposes [17]. Due to computerisation of the modern power plants, huge amounts of data are readily available to be processed by machine learning algorithms. Therefore, it becomes mandatory to figure out what parameters can influence the detection of faults in the said machinery.

The key step is to identify the features that separate the faulty machines from the non-faulty ones. For example, faulty machines may have abnormal motor oscillations, may have unnatural sounds than non-faulty ones, cracks in the body etc. These abnormalities are then processed as images, sound data, frequency signals, etc. which are then converted to a suitable mathematical format. After data processing then it is possible that some features may have values that may influence the model parameters regardless of the impact of that feature; therefore to avoid this problem, the data is normalised to bring all the features to a similar level. The feature selection is applied to rule out the irrelevant or redundant features that don't play any role or have a very low impact on the training of the predictive model. This is usually done with help of feature selection methods, like correlation analysis, decision trees, linear discriminant analysis, etc. Due to the processing of such complex data like images, and sound frequency signals, the dimension of the data may become extremely large and hence increase the training complexity. Therefore, to avoid this problem, data reduction may be applied [see fig. 4.8].

The data after going through the above process is then used for training the support vector machine algorithm with the help of a kernel function. The radial basis function is usually applied to train the kernel SVM model as it has some desirable properties. The radial basis function, in general, is

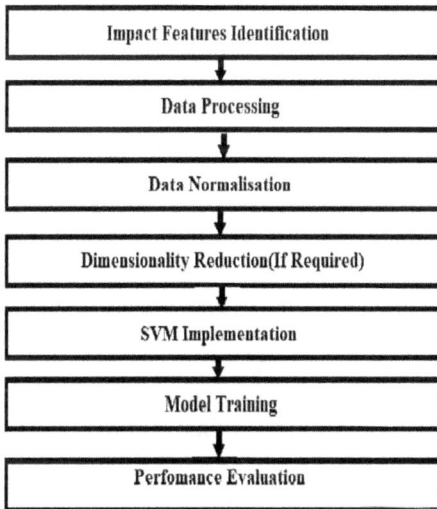

Impact Features Identification

Data Processing

Data Normalisation

Dimensionality Reduction(If Required)

SVM Implementation

Model Training

Perfomance Evaluation

Figure 4.8 Shows the key steps involved in the implementation of the support vector machine algorithm for diagnosing the faults in the steam turbine generator

a generalised version of most kernel functions like the quadratic, and cubic kernel functions. Further, as it always projects the data in the range of 0 to1, a lot of difficulties related to large numbers. Since the SVM model is highly sensitive to the hyperparameters, the radial basis function only has to tune the two parameters s and α, thereby, not adding much to the complexity of the classifier. After tuning the SVM with optimal hyperparameters, the model is again trained with this set of hyperparameters to generate optimal results. The model is then implemented with performance evaluation methods like accuracy, robustness, scalability with other data, and so on.

4.5.2 Product quality and management

One of the most highly complicated and technologically challenging items that can be manufactured is the semiconductor. Failing to meet the required quality can lead to crude performance and thus lead to problems in devices that use semiconductors like graphics cards, processors, etc. Therefore, it becomes extremely necessary to maintain quality standards in the manufacturing of semiconductors. The process of manufacturing the semiconductor involves a cycle of a complicated and prolonged process of generating wafers involving a large number of procedures which requires various sensors to be installed on the equipment. The sensors installed convert the signals and other information into programmable data for fault detection and classification purposes.

4.5.2.1 Semi-conductor quality control

In the semiconductor quality control process, the wafer of the semiconductor must be handled with utmost care [8, 14]. The quality control process of this wafer is usually classified into two categories, the work-in-process test and the control wafer testing machine. The quality control methods deployed in the quality control of semiconductors are:

- **Inspection:** The defects observed in the wafers of the semiconductors are inspected by the workers or the experts available on site. This is primarily achieved by observing the wafers by just observing the wafer with the naked eye to a microscopic level. Some sampling methods are applied to observe the wafer qualities which yield both quantitative as well as qualitative methods.
- **Offline measuring machines testing:** A dummy wafer is used to simulate the results of machine processes. As soon as a wafer is manufactured, it is directly forwarded to the measuring machines. This uses the statistical process control (SPC) with some quantitative data. This data is handled by some operator and the counting function can be handled by computer integrated manufacturing (CIM).

- **Defect analysis:** The defects found on the surface of the wafer are then analysed with the help of some defect analysing instruments to learn the depth of the defects found.
- **Wafer acceptance test:** The electronic circuits built on the wafer are assigned five points for inspection and each of these five inspection points have about one fifth of the area for the die test.
- **Die test:** The testing machines test the dies at highest possible resolution for in depth analysis. This, even though it ensures maximum quality, is extremely time consuming.

Every moment the technology involved in semiconductors sees a sharp rise in innovation. Due to the ever-increasing levels of innovation, the semiconductors have become extremely complicated and hence the time required to detect the faults and further localise them has increased exponentially.

In order to apply machine learning in outlier detection of semi-conductors, we apply artificial neural networks and to implement this, the parameters of the feedforward network must be established. The hyperparameters of the model, like, number of hidden layers, learning rate, etc. must also be figured using experimental results [see fig. 4.9].

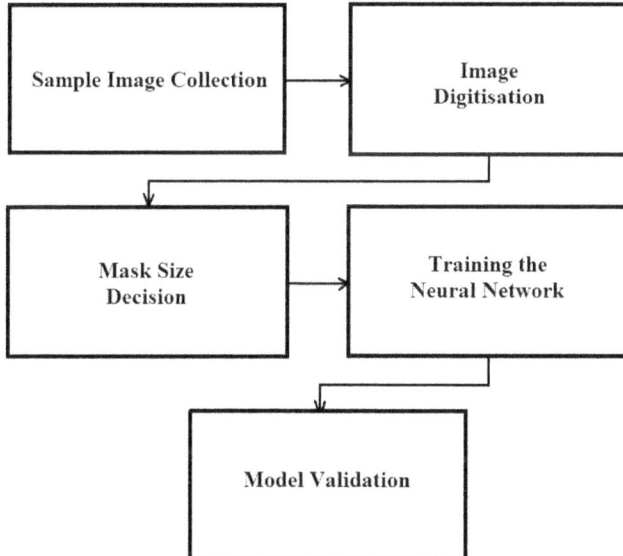

Figure 4.9 Illustrates the five major processes that are involved in the quality control process of semiconductors by artificial neural networks as explained

Sample image collection: It is essential to gain adequate amounts of data for application of the neural networks to the quality control process of semi-conductors. To fulfil this, sample data of both the defective and non-defective semiconductor wafer dies are collected. The images of the defective dies must contain the most frequently occurring defects in the wafers and must be labelled appropriately.

Image digitisation: The collected images are then digitised by placing the images on a piece of black paper and images are scanned to a system. Then the orientation of the images is adjusted accordingly and finally the black background is removed by the means of some photoshopping software. This is completely manual and hence the most time-consuming parts of the process and can be automated in future.

Mask size decision: With the image digitisation process the images are converted to grayscale. If Each pixel of the digitised image is considered as one input for the nodes in the input layer, the application of input nodes becomes rather tedious and the size of the network becomes extremely huge. To get rid of this problem, a matrix of size $N_1 \times N_2$ of pixels is considered in which N_1 is the number of pixels in rectangular width and N_2 is the number of pixels in rectangular length. Inside this mask the average value of intensities of pixels is considered. When $N_1 = N_2 = 1$, we consider each pixel as an input node. Therefore, as big the size of the mask is, the smaller will be the neural network. A large size of the mask may hamper the quality of the neural network generated. The mask size creates a trade-off between the accuracy of the network and the amount of time required to process the information. This is one of the hyper-parameters of the neural network which needs to be experimented with. One of the most popular approaches to figure out a good mask size is to take a sequence of mask sizes and then apply a neural network on the data images and choose the one that has the best performance. The masked data samples are then converted to training and testing data

Training the neural network: After generating the input data with varying mask sizes the neural network is trained. Usually, one hidden layer is preferred but the number of hidden layers may be increased to get better performance. The number of hidden layer nodes is obtained via experimental results. The learning and momentum rates are initially fixed at some predefined value. The loss function is defined to be a binary cross entropy function. The values of the learning and momentum rates are adjusted accordingly as the loss gets stable.

Model validation: To evaluate the performance of the model in terms of both efficiency and effectiveness, the neural network is fed the testing data, i.e., masked images that were split from the original dataset to form the testing dataset.

After generating a perfect neural network model, the need to validate the algorithm as one of the best ones arises. It is compared with various other algorithms like the Bayesian networks, hidden Markov Models, SVM, random forest classification, etc. The key factors to evaluate the performance of the models are inspection area accuracy, speed, stability, cost, ease of operation, and flexibility.

4.5.2.2 Implementation of the artificial neural networks for semiconductor fault diagnostics

To diagnose the faults in the semiconductors, the entire assembly line must be implemented with the ANN model. Various hardware components like scanning electron microscopes (SEM), conveyor belts, and die pickers are involved in the automation process for anomaly/fault detection.

The process of transfer of the wafer from the sawing process to the point of inspection has a conceptualised automatic inspection system. The SEM picks up an image from the wafer which is digitised with an appropriate mask size and then is fed to the neural network that will be trained with a backpropagation algorithm. If the neural network deems the die as not faulty then it is moved forward, if the die seems to have some faults, then the next die image is scanned and the process continues until all the dies have been inspected. The wafers with defective dies are discarded and the process goes on until all the wafers have been classified [18].

4.6 CHALLENGES IN APPLYING MACHINE LEARNING IN MANUFACTURING

One of the toughest tasks in applying machine learning in production and manufacturing is the absence of guidelines specified by the industries. This prevents companies from applying artificial intelligence in the way they want to address the issues they are facing [22].

Even though data are abundant at the disposal of the companies, it, however, does not imply that this data will always be relevant. The problem that this data possess, is that no matter how well the model has been built, if this data is fed to the model, it may end up delivering disastrous results and would lead to discontinuities in the manufacturing processes.

Applying machine learning to manufacturing processes poses the problem of scalability. More often than not, some reasons lead to very low scalability of the machine learning models, i.e., they can only perform well under some special circumstances and can't be applied to various other problems faced by the industries. This is often a major concern for data scientists and analysts and further modifications are required in the model to be able to adapt to other problems which lead to increased cost and time in deploying the models.

Apart from the model itself, the training data may also pose the problem of non-representation, i.e., it may not be able to represent new cases that we want our model to generalise to. Using this data to train the model would lead to reduced performance, and low accuracy and may lead to biases towards a specific type of class [21].

The challenges faced in fault detection and classification in semiconductor manufacturing include building a classifier which can detect the abnormalities in the wafer with a high degree of accuracy, identify the major SVIDs, i.e., status variable identification, processing time, and steps for classification in a high dimensional feature space. To efficiently analyse the SVID data, artificial neural networks can be applied. ANNs are endowed with features like possess learning, fault tolerance, and parallel computing. If these functions are applied correctly, ANN can develop a very accurate predictive model for detecting outliers thus improving the process of manufacturing the semiconductors which in turn would improve and enhance the quality of the semiconductors manufactured [16].

REFERENCES

1. Zaki, M.J., Meira, W., *Data Mining and Analysis:Fundamental Concepts and Algorithms*, Cambridge University Press, 2016.
2. Hastie, T., Tibshirani, R., Friedman, J., *The Elements of Statistical Learning: Data Mining, Inference, and Prediction*. Springer, 2nd Edition, 2017.
3. Wuest, T., Wiemer, D., Irgens, C., Thauben, K., Machine learning in manufacturing: Advantages, challenges, and applications.*Production and Manufacturing Research*4 (2016) 23–45.
4. Shyun, H.J., Eom, D.H., Kim, S.S., One-class support vector machines—An application in machine fault detection and classification. *Computers and Industrial Engineering*48 (2005) 395–408.
5. Seongmin, H., Lee, J.H., *Fault Detection and Classification Using Artificial Neural Networks*. IFAC International Federation of Automatic Control, 51 (2018) 470–475.
6. Mitchell, T.M., *Machine Learning*. McGraw-Hill Science/Engineering/Math, 1997.

7. Chen, K.Y., Chen, L.S., Chen, M.C., Lee, C.L., Using SVM based method for equipment faultdetection in a thermal power plant, Computers in Industry 62 (2011) 42–50.

8. Munirathinam, S., Ramadoss, B., Predictive models for equipment fault detection in the semiconductor manufacturing process. *IACSIT International Journal of Engineering and Technology*8(4) (August 2016) 273–285.

9. https://www.twi-global.com/technical-knowledge/faqs/faq-what-is -manufacturing.

10. Vaidya, S., Ambad, P., Bhosle, S., Industry 4.0 - A glimpse, 2nd international conference on materials manufacturing and design engineering. *Procedia Manufacturing*20 (2018) 233–238.

11. https://www.itconvergence.com/blog/6-benefits-of-machine-learning-in -manufacturing.

12. https://marutitech.com/machine-learning-in-supply-chain/#What_is_ Machine_Learning.

13. Zhong, S. et al., Machine learning new ideas and tools in environmental science and engineering. *Environmental Science Technology*55(19) (2021) 12741–12754.

14. Bajic, B., Cosic, I., Lazarevic, M., Sremcev, N., Rikalovic, A., Machine learning techniques for smart manufacturing: Applications and challenges in industry 4.0, 9[th]International Scientific and Expert Conference TEAM 2018(2018)

15. Stanisavljevic, D., Spitzer, M., A review of related work on machine learning in semiconductor manufacturing and assembly lines, SAMI@ iKNOW(2016)

16. Xu, X., Zhang, X., Fault detection for turbine engine disk based on adaptive weighted one-class support vector machine. *Journal of Electrical and Computer Engineering* (2020).

17. Su, C.T., Yangb, T., Ke, C.M., A neural-network approach for semiconductor wafer post-sawing inspection,*IEEE Transactions on Semiconductor Manufacturing*15 (2002) 260–266.

18. Ademujimi, T., Brundage, M.P., Prabhu, V.V., A review of current machine learning techniques in manufacturing diagnosis,*IFIP Advances in Information and Communication Technology* (2017) 407–415.

19. Fan, S.K.S., Hsu, C.Y., Tsai, D.M., He, F., Cheng, C., Data driven approach for fault detection and diagnostic in semiconductor manufacturing,*IEEE Transactions on Automation Science and Engineering*17(2020) 1925–1936.

20. https://www.analyticsvidhya.com/blog/2021/06/5-challenges-of-machine -learning/

21. Mayr, A., et al., Machine learning in production - potentials, challenges and exemplary applications, 7th CIRP global web conference, "Towards shifted production value stream patterns through inference of data, models and technology."

22. Nikolic, B., Ignjatic, J., Suzic, N., Stevanov, B., Rikalovic, A., Predictive manufacturing systems in industry 4.0: Trends, benefits and challenges. Proceedings of the 28th DAAAM international symposium on intelligent manufacturing and automation (2017) 796–802.

23. Kang, H.S. et al., Smart manufacturing: Past research, present findings, and future directions. *International Journal of Precision Engineering and Manufacturing – Green Technology*3(1) (2016) 111–128.
24. Lee, J., Kao, H.A., Yang, S., Service innovation and smart analytics for industry 4.0 and big data environment. *Procedia CIRP*16 (2014) 3–8.
25. https://www.analyticsvidhya.com/blog/2021/05/support-vector-machines/.

Chapter 5

Advancement of machine learning and image processing in material science

Ayush Pratap and Neha Sardana

5.1 INTRODUCTION

Scientists can interpret the material data but by integration of machine learning the interpretation gets accelerated with better accuracy. This intervention of machine learning in material science has changed the research perspective drastically [1]. Previously tremendous time and effort were required in the experimental method, for the development of new material. Recently, the material science research has delved into ML to reduce time and increase accuracy [2]. Thus, for analysis or interpretation of research data it becomes easy by using a simple ML model, although it is complex to decide which model to select for what purpose. The model selection has been explained in this chapter thoroughly [3]. ML is a technique or a set of procedures that uses a computer system in which the user does not require proficient knowledge of programming. ML is subdivided into two broad categories: Supervised (where a large amount of label data is to be used) [4] and unsupervised learning (where there is a lack of label data) [5]. The technique of supervised, unsupervised, semi-supervised, and different ML models has been discussed in this chapter [6]. The human has evolved with various new evolutions in technology and material. Advanced material is a part of that evolution that helps mankind and makes the foundation of the new emerging industry. Traditionally the time which is required for advanced material from lab to application was quite long, approximately 10 to 20 years. To accelerate the discovery of material, Materials Genome Initiative is made which is an infrastructure of materials data [7]. Image processing [8] is a classification technique in which the deployment of the image is being done with some short of pre-work [9]. Thus, it is better to build an ML model that can be easily deployed and can be used for testing various intelligent systems around it [10]. The most important part of the ML process is to select an algorithm or model. This model can only be selected efficiently if you understand your data correctly and know the relationship with the problem you want to solve. Each algorithm has been designed for a different kind of work [11]. ML helps in identifying important nonlinear multivariable relationships for a range of material classes. These

DOI: 10.1201/9781003246466-5

classes include metals, alloys, ceramics, composite, polymer, 2D materials, organic inorganic hybrids, etc. Sometimes we get a misleading conclusion due to unorganized data which lead to models breaking down [12]. The various approach for material prediction using ML has been done, which include estimation of energetics [13]–[15], bandgaps [16], [17], glass transition temperatures [18], mechanical and elastic properties [19], the strength of additive manufacturing component [20], thermal conductivity [21], [22], Microstructure recognition [23]. It has seen a thrust in these areas which has grown in the last decade [24].

In this work, the historical development of ML has been narrated with a timeline, and how material science has adopted it in its domain has also been elaborated. After the assessment of various ML applications in material science and connected fields, the aim is on the ML model selection and image processing technique along with the problem which is associated with the material science research in image processing. The chapter focuses on various software and platforms used for image processing and also classified the different ML models based on their similarities. Handpicked applications from the latest studies are highlighted. The problem associated with small datasets is pointed out and the solution to it has also been explained [25]. At last, various platforms such as MDF [26], MGI [7], DLhub [26]–[28] are also discussed to make data collection easy. Finally, a SWOT analysis of image processing in material science has been done through which it becomes easy to understand the positive and negative perspectives of this technology.

Throughout the chapter, it is presumed that the readers are familiar with the basics of ML and its algorithm applied in the region of material science. The basics of python, various libraries like TensorFlow [29], sci-kit learn [30], and modules like Kera's are not covered but could be found elsewhere [31].

5.2 HISTORICAL EVOLUTION OF MACHINE LEARNING IN MATERIAL SCIENCE

Machine learning is involved in every aspect of life like an autonomous car, creating match reports, predicting natural disasters, or finding the terrorist suspect. ML is the most commonly heard buzzword in the recent past. The evolution of humans has been taken with different materials and technology. This evolution has led to the foundation of modern industry. Neither cultural transmission nor toolmaking is unique to humans [32]. Material is a basic building block of any industrial system. Thus, from prehistoric times there is always a quest to find new material for advanced technology. But traditionally more time and cost were required for the development of novel material. Thus, the incorporation of ML in the development of material has reduced time and resources. The flow diagram on human evolution with the material is shown in Figure 5.1.

The foundation of Industry 4.0 is the data. This data is generated with the help of various smart sensors, devices, and machines during the time of production [33]. This data is used to increase the efficiency of the manufacturing sector by actionable intelligence predicted by ML algorithms. Using an ML algorithm provides a predictive insight to make intelligent decisions in different manufacturing tasks like process optimization, supply chain management, intelligent and continuous inspection, predictive maintenance, quality improvement, and task scheduling. Thus the productivity and total manufacturing has been optimized by the use of a data-driven approach [34]. The traditional boundaries of materials development and industry are shifting towards computing, digital technology, and image processing. Self-learning solutions and capabilities decrease the overall cost of material development and production which lead to the foundation of Industry 4.0 [35]. The world we are living in today is generating an ample amount of data and the availability of this data is increasing the quest for using machine learning solutions and practices [36].

The origin of AI was in 1950 when Alan Turing, a computer expert, communicated the article "can machines think," a hypothesis propounded by him that a machine can succeed in convincing humans. This was called the Turing test. The psychologist Frank Rosenblatt from Cornell University developed machine learning in the modern sense in 1957. He was the one who laid the foundation stone of neural networks for computers, today called perceptron. The perceptron algorithm was able to categorize the image input into one of two groups. The whole idea is based on the work of the human nervous system. Bernard Widerow and Marcian Hoff created two NN known as Adeline (for binary pattern detection) and Madeline (to

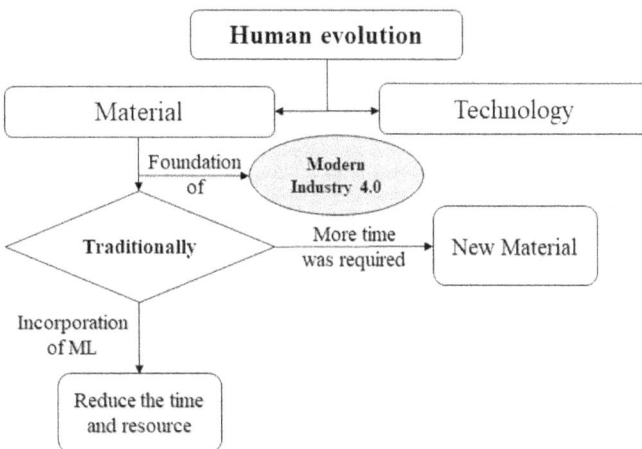

Figure 5.1 Material's evolution with time and technology

remove echo from phone lines) in 1959. In 1967, a neural network algorithm was written which later allowed the computer to recognize the pattern. The idea of explanation-based learning was introduced in which a computer discards information by following some general rules in 1981. In the 1990s ML shifted to a database approach which is easier as compared to a knowledge-based approach. In recent times different computer programs have been developed that can beat humans in various mind games. In 2002 IBM Watson beat two human champions in the game of jeopardy. In 2016 Alpha go was the first computer program to beat human professionals. The 21st century witnessed the real development of machine learning at the ground level. Many business ventures have delved into various ML projects, like Google Brain, Alex Net, Deep Face, Deep Mind, Open AI, Amazon ML Platform, and Resnet. Amazon, Netflix, Google, Salesforce, and IBM are dominating the IT industry with ML. A summary of the historical development in ML for various works is listed in Table 5.1.

Table 5.1 Historical development of machine learning

Year	Work	Reference
1958	Perceptron was created by Frank Rosenblatt. The building block of Artificial Neural Networks (ANN).	[38]
1981	Gerald Dejong proposed Explanation-Based Learning (EBL). To analyze training data and make rules to discard unimportant data.	[39]
1985	Terry Sejnowski developedNetTalk.to pronounce English words as pronounced by children.	[40]
1986	A new architecture of neural networks was developed by Geoffery Hinton. He coined the term deep learning. This uses multiple layers of neurons for learning.	[41]
1990	ML has moved from knowledge-driven to a data-driven approach. Now the machine can analyze the data and can make important decisions from it.	[42]
2011	Jeff Dean developed Google Brain. A deep neural network for videos and images.	[43]
2014	The "Deep Face" was invented by Facebook which was based on deep neural networks.	[44]
2015	"Distributed Machine Learning Toolkit." To distribute the problems of machine learning to multiple computers parallelly for a common solution.	[45]
2016	The AlphaGo program was developed by Google that can beat a professional human.	[46]
2017	Google vizier a de facto parameter tuning engine developed at google.	[47]
2020	Materials Simulation Toolkit for Machine Learning (MAST-ML) was developed, which is a software package designed for accelerating material science research.	[48]

ML has acquired a platform in data science logarithmically in the recent past as the quality of data we produce continues to grow, so we must have the computational aptitude and ability to process and analyze it [37].

Using machine learning in material science or any field required some basic steps which are mentioned in Figure 5.2 (a). Thus, the very basic step is the collection of data from any work environment. While collecting the data there is always noise. It should be removed from the data and further visualized after cleaning. This cleaned data is further used in the model and it is evaluated based on error similarity or probability. The different machine-learning approach is demonstrated in Figure 5.2 (b). If the accuracy is within the specified limit the model is deployed for the desired task for which it is prepared.

5.3 MACHINE LEARNING ALGORITHM

At its core, data science helps us make sense of the massive world of information all around us, a world that's far too complex to study directly by ourselves. Data is the record of everything that's going on and what we should learn from it. The real value of all this information is what it means. Machine learning helps us discover patterns in data, which is where meaning lives. When we can see what the data means, we can make predictions.

Machine learning algorithms and machine learning models are two distinct sides of the same coin. At one phase it can be said that ML algorithms are a procedure that is implemented on code and the models are output that we get by that algorithm. There is no such algorithm that can be used for

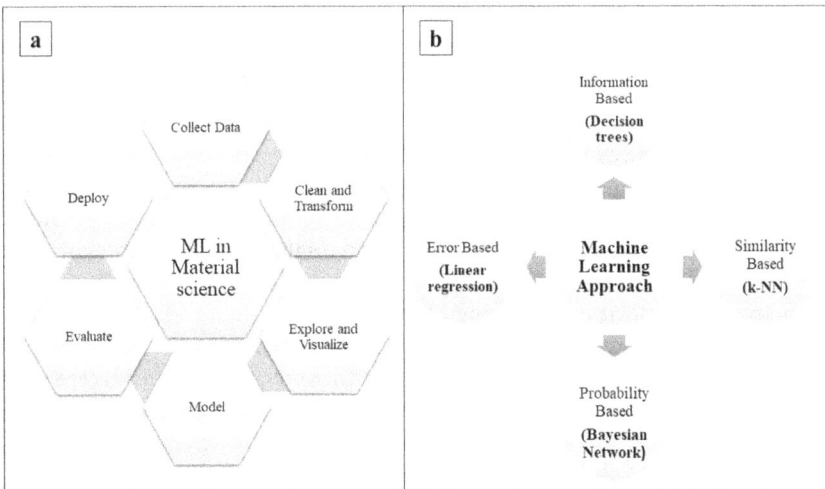

Figure 5.2 (a) Material science research through ML; (b) Various machine learning approaches

all datasets. Thus, we have to select a model and verify it for our datasets based on its accuracy and performance [49]. Our brain works for every new situation and what it has learned from the past. This is how the decision-making of the brain is improved progressively by interacting with the various environments and surroundings [50]. There are various kinds of ML models; for the time being a neural architecture depends on the application required taking into account the features to be detected [51]. Here the fishbone diagram of machine learning to a given data is designed in Figure 5.3. The various algorithms can be chosen for a given data for a desired output solution or interpretation.

5.3.1 Supervised learning

There is a type of ML model in which the machine is trained using an ample amount of labelled data based on which the machine will predict unlabelled data in the future. This kind of algorithm is known as supervised learning. Here the relation is built between the input and output vector which is based on a large amount of training data [4]. In Figure 5.4 (a), the image dataset of stars and moon were classified. It works in a similar way of training the algorithm with labelled data of stars and moon and finally testing it with unlabelled data of the same image. A few examples of supervised machine learning in the material science domain are Naive Bayes [52], decision trees [53], [54], linear regression, convolutional neural network (CNN) [20], [55]–[58], genetic programming [59], long short term memory [60], artificial neural network (ANN) [61], particle swarm algorithm [62], k-nearest neighbour (KNN) [63], radial basis function [64], Siamese NN [65], and support vector machine (SVM) [66], [67].

5.3.2 Unsupervised learning

There is also a type of ML algorithm in which the datasets are not labelled and the sole purpose of the model is to find out a specific pattern in the given data points. This kind of algorithm is known as an unsupervised learning model. The most common example is clustering where the data is grouped by a similar feature. From Figure 5.4 (b) unlabelled images of stars and the moon were clustered in a specific group. This could be achieved by identifying a specific pattern in the shape of star and moon datasets. Another one is principal component analysis where the algorithm itself finds a way to reduce the dataset, by keeping the important one and discarding the rest. Various examples of unsupervised learning used in the field of material science are K-means clustering [68], [69], self-organizing map (SOM) [70], [71], and restricted Boltzmann machine [72].

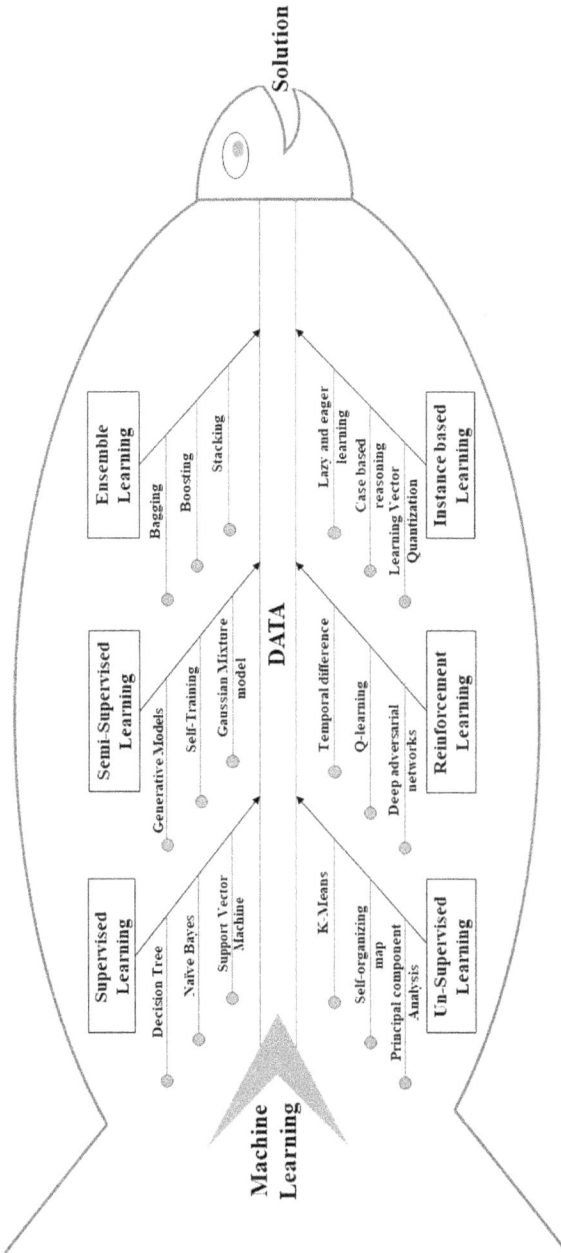

Figure 5.3 Fishbone diagram of machine learning

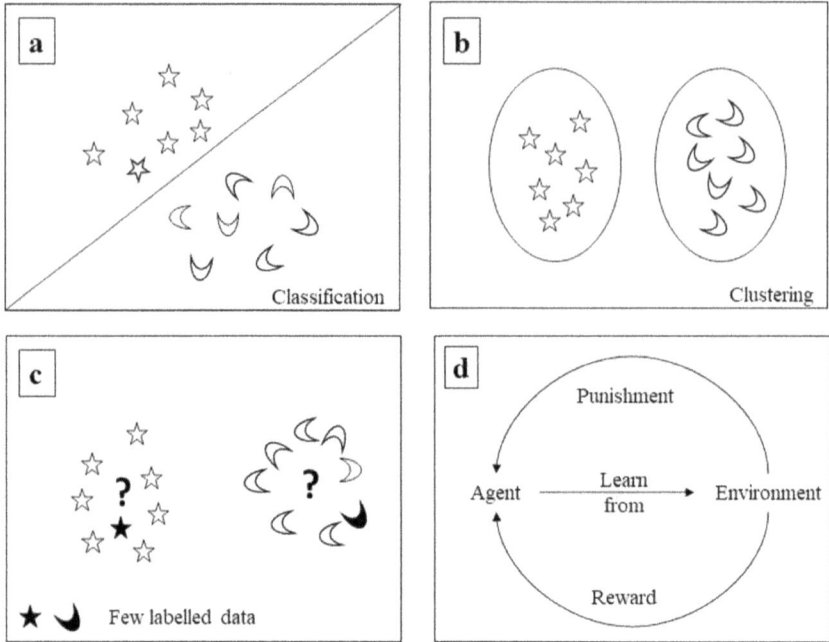

Figure 5.4 (a) Supervised; (b) Unsupervised; (c) Semi-supervised; (d) Reinforced

5.3.3 Semi-supervised learning

A synergistic approach of using both supervised and unsupervised learning algorithms together is known as semi-supervised learning. Labelling an ample amount of a dataset is too costly, thus this model is a mixture of labelled and unlabelled data use. Through this, we can harness an ample amount of data from various use cases with some amount of labelled dataset [73]. The existence of some labelled data makes this model perform better than unsupervised learning. Thus, the cost and time to train the model get reduced as compared to supervised machine learning. From Figure 5.4 (c) it is shown that there is a little amount of labelled data present in the dataset of unlabelled data. That combination of labelled and unlabelled data is used for training a model. Various examples of semi-supervised algorithms are graph-based methods, generative models, the Gaussian Mixture model [74], and low-density separation.

5.3.4 Reinforced learning

A subfield of ML is where a model learns by itself gradually by using trial and error or by experience. This kind of algorithm is known as reinforcement learning. The most crucial part of reinforced learning (RL) is that a model learns good behaviour by itself. Thus, for the RL algorithm complete information about the environment is not required, rather it collects information from the environment just by interacting with it [75]. Reinforced learning tells us about the dataset and whether it is correct or not. This good or bad data is decided by the algorithm by interacting with the environment. The model interacts with the environment and gets a reward or punishment. This positive and negative feedback reinforces the model; thus, the algorithm is known as reinforcement learning. For instance, a child has been attracted to candlelight and approaches it. Thus, his action will decide the punishment and reward for him and give him an experience for the future. This can be well understood by Figure 5.5. Various examples of reinforcement learning algorithms are the temporal difference and deep adversarial networks [76]–[78], and Q-learning [79].

5.3.5 Ensemble learning

The use of multiple algorithms together for a specific task is known as ensemble learning. This enhances the predictive performance of the algorithm which cannot be achieved by a single algorithm. It follows the principle of "the wisdom of the crowd," which means that when a group of people wants to solve a problem then it becomes easy to solve a task.

Although ensemble learning is a powerful method, it has some drawbacks. The first one is that while ensembling, more time and resources are required to train multiple ML models. Secondly, a large amount of memory is used as more than one algorithm has to be trained simultaneously. Ensemble

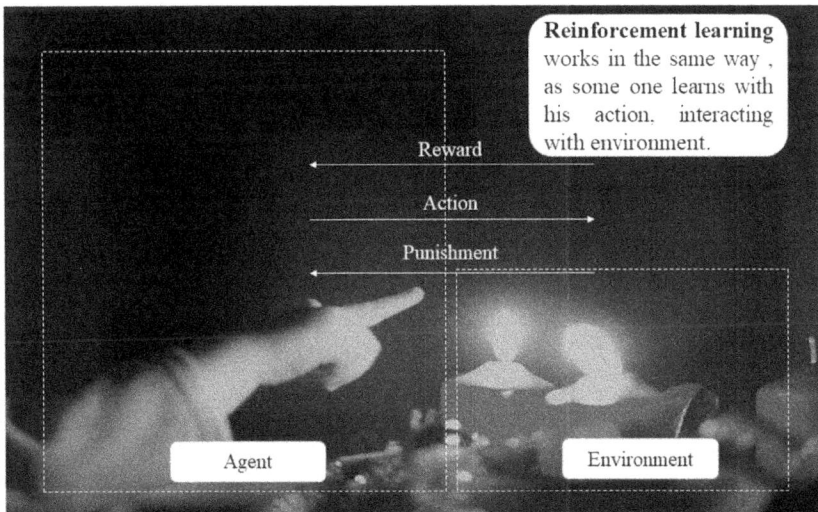

Figure 5.5 Reinforced learning

techniques have been further classified as Bagging and Boosting. Both algorithms use voting and combine a model of the same type. In Bagging the individual models are built separately with equal weights assigned to each model. Rather in Boosting, the model works in series and the next model is dependent upon the work of the previous model. The weights are given to the model as per performance [80]. There are plenty of other models and submodels in ML. The various kinds of models are classified based on similarities and are listed in Table 5.2.

5.4 IMAGE PROCESSING TECHNIQUE

ML can interpret in the same way as our brains do this when talking about the image data processing that we can see in our day-to-day life while shopping from Flipkart or unlocking the phone with a face scan or using Instagram. Humans can interpret the acquired visual information they have in the long run of life. Similar to this, the ML algorithm also works with a large amount of image data and visual scenes [8]. Machine learning is an algorithm in which the model learns in three basic steps [90], which include task performance and experience. The complete flow diagram for image processing is shown in Figure 5.6.

No where is it specified where to use which model, thus the decision for the model section for any given dataset is quite important. Today in the digital world there is an ample amount of available data. Segregating the correct data for analysis or prediction is itself a great task. Any data can be of two categories, one is correlated and the other is redundant. Considering the material perspective if the data is correlated then it can be used directly by reducing its dimension. Now the question arises whether the data has a category or not, if yes then that should be labelled for further processing or model selection (which is the most crucial part of any ML work). The flowchart for the model selection is shown in Figure 5.7. Classification and clustering are two important techniques that can be performed on the labelled data. The classification is an algorithm through which the model is trained with a labelled vector and a new unlabelled vector is used to check the accuracy of the model. The classification can be of two types one is binary (e.g., cat and dog image dataset) and another is multiclass (different weather dataset). Additionally, clustering is a method through which unlabelled data can be used for interpretation. Herein a decision boundary has been made which divides the data into different similar groups. Now if the data does not have a category, the regression algorithm can be used for quantitative analysis.

Recently many industries in the manufacturing process are interested in automation by using AI which also reduces the cost of designing the material [91]. Digital image processing (DIP) is a computational method used

Table 5.2 The algorithm based on similarity

Algorithm	Basic	Type	Reference
Regression	Regression works on the measure of an error made by the model in any prediction work. It is a statistical machine-learning method	• Linear and Logistic Regression • Locally Estimated Scatterplot Smoothing (LOESS) • Multivariate Adaptive Regression Splines (MARS)	[81]
Instance-based	It is a prob of decision-making. In this method, the database is built to compare new data to find some similarities to find the best match	• Support Vector Machines (SVM) • k-Nearest Neighbour (kNN) • Self-Organizing Map (SOM) • Learning Vector Quantization (LVQ)	[82]
Regularization	It favors a simpler model which is better for regularizing	• Least Absolute Shrinkage and Selection Operator (LASSO) • Ridge Regression and Elastic Net • Least-Angle Regression (LARS)	[83]
Decision tree	This method is often fast and accurate. This is used very often in ML.	• Conditional Decision Trees • Decision Stump • Classification and Regression Tree (CART) • Iterative Dichotomizer 3 (ID3) • Chi-squared Automatic Interaction Detection (CHAID)	[84][54] [53]
Bayesian algorithms	This applies Bayes' Theorem.	• Bayesian Belief Network (BBN) • Bayesian Network (BN) • Naive Bayes • Gaussian and Multinomial Naive Bayes	[85]
Clustering	It describes a class of problems and methods.	• Hierarchical Clustering • Expectation Maximization (EM) • k-Means • k-Medians	[74][86]

(Continued)

Table 5.2 (Continued)

Algorithm	Basic	Type	Reference
Association rule	It explains observed relationships between variables in data.	• Eclat algorithm • A priori algorithm	[87]
Artificial neural network	This is a model that is inspired by the structure or function of the human brain.	• Perceptron • Multilayer Perceptron's (MLP) • Back-Propagation • Stochastic Gradient Descent	[61]
Deep learning	This is a new version of ANN that can use an ample amount of data with cheap computation.	• Convolutional Neural Network (CNN)\ • Deep Belief Networks (DBN) • Long Short-Term Memory Networks (LSTMs) • Stacked Auto-Encoders • Recurrent Neural Networks (RNNs)	[88]
Dimensionality reduction	This also works as a clustering method but in an unsupervised manner.	• Partial Least Squares Regression (PLSR) • Principal Component Regression (PCR) • Multidimensional Scaling (MDS) • Principal Component Analysis (PCA) • Linear Discriminant Analysis (LDA)	[89]

to help many areas with decision-making by extracting visual information from images [11]. A digital image is a matrix composed of pixels whose positions are given by x and y values, that is, a mathematical function f (x, y) [4]. A DIP system can be divided into the following steps: acquisition, pre-processing, segmentation, feature extraction, recognition, and interpretation. The acquisition is responsible for converting a real scene captured by a digital camera or other devices into a digital image. In pre-processing some techniques for digital image correction are applied so that the image can be better processed in the following steps as shown in Figure 5.6. Segmentation seeks to split an image into regions or objects of interest. This process is guided by object features such as shape, texture, and colour. In the feature extraction step is generated a database with characteristics of detected objects in the segmentation step, such as size, area, and shape. Finally, in the recognition and interpretation step, objects detected in the

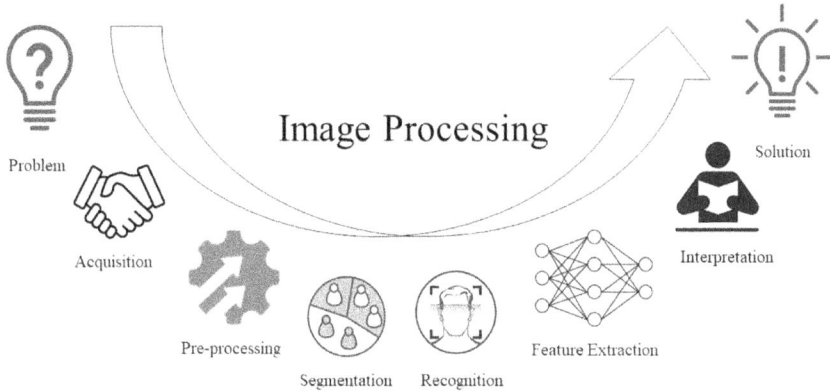

Figure 5.6 Image processing flow diagram

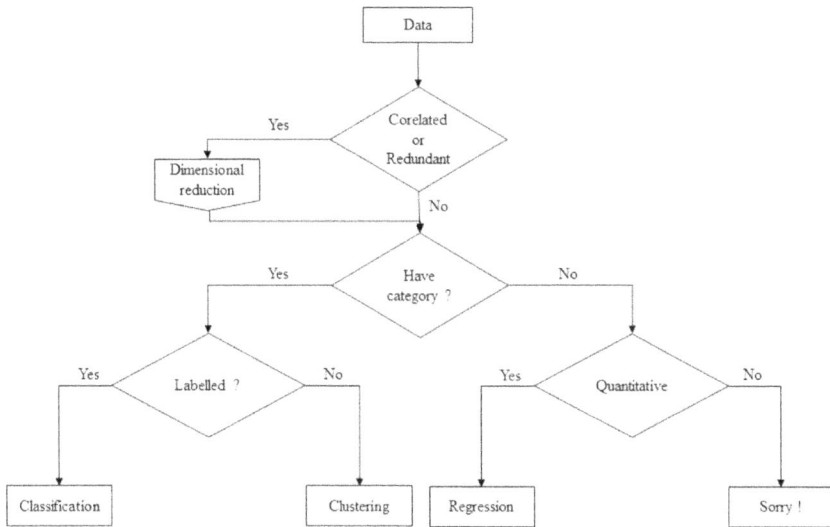

Figure 5.7 Flowchart for model selection

segmentation stage, together with their characteristics, are classified and interpreted according to the desired solution [92].

For the software development solution for DIP, we need a programming language that would support the technique which is necessary for image processing. Python [93] is an object-oriented language, with an excellent learning curve, which supports the development of algorithms for image processing. Furthermore, Python is a language with compact syntax, simple, and multi-platform, which facilitates the use of its programs in automation solutions. SimpleCV is a specific framework of digital image processing for use with Python. A tensor flow module can also be used for making the analysis simpler. The various image processing libraries and frameworks are listed in Table 5.3.

Table 5.3 Image processing library and framework

Library/ framework	Description	Features	type	Developed by	Year
OpenCV	It is a python library used to solve computer vision problems.	It has a big library for image processing which can work on both images and videos. It also has a Java API Extension.	Open Source	Intel	2000
TensorFlow	Most popular ML framework.	It can work on various multiple parallel processors and have different ML and NN algorithms. It is cross-platform.	Open Source	Google	2015
PyTorch	It works mainly on python as compared to any other framework.	It has big community support and can work with GPUs.	Open Source	Facebook	2016
Caffe	a deep learning framework	It is C++ Based deep learning framework that has an expressive architecture.	Open Source	Berkeley AI Research (BAIR)	2017
Emgu CV	Real-time computer vision library.	Cross-platform	Open Source	Apache software foundation.	2017
MATLAB Image Processing Toolbox:	Image Processing Toolbox apps.	Can segment image data and can use large datasets.		MATLAB	
Apache Marvin-AI	AI platform that helps deliver complex solutions with low-latency,	Multi-threaded image processing can be done by using this framework.	Open Source	A merger of the 3 leading e-commerce companies in Brazil	2006

5.5 PROBLEMS AND SOLUTIONS ASSOCIATED WITH IMAGE PROCESSING

The technical analysis of an image by using a complex algorithm is known as image processing. But for analysing image data you must have an ample amount of dataset. Image processing is currently being used in different sectors away from health care, defence, agriculture, etc. For instance, considering material science research the dataset is quite small compared to any other sector. Thus, the detailed analysis of the smaller dataset and its solution is well described in this work, keeping it much more relevant to the materials perspective.

5.5.1 Smaller datasets

When people hear about ML, the very first thing that comes into mind is big data. Most of the research fields including material science research are data-hungry. Yet, as compared to other fields the data in material science is smaller and random which affects the ML model applied in this area [94]. Recently the power of machine learning has been increased in this research with the help of various developed tools and initiatives. Some of the practical tools are data augmentation and transfer learning. Also, an initiative like MGI and DL hub has helped very well. The prediction for screening of material is most promising in the use of machine learning [95]. The various problem associated with small dataset are listed below:

- **Overfitting:** This is the most common problem faced when using a small dataset. This is a modelling error that occurs when our model function corresponds too closely to a given dataset.
- **Outliers:** There is always some data in the dataset which is eccentric from the rest of the data. That data is known as outliers. Thus, handling that data is quite important, but for big datasets, few outliers can be lived. But for small datasets, even a few outliers can give a significant error to the model.
- **Train and test data:** For a good model design the data must be split into training and test data. This test data, sometimes called holdout data, is used for cross-validation of the model. Thus, having a small dataset one does not have this opulence, and even if one does, the number of holdout data will be very small to give a performance evaluation of the model.
- **Measurement error:** The data which is gathered from the real world has some measurement error associated with it. When the data is big then this error has no significance but in the case of a smaller dataset this error can adversely affect the model.

- **Missing values:** Gathering data from various sensors is sometimes not so accurate. There might be a time when some of the data is missed while recording. These missing values at a small scale adversely affect the model.

This issue of small dataset can be solved by a few solutions which are mentioned below:

- **Domain expertise:** Small data does not offer the luxury of testing different models hence an expert from the specific domain counts more.
- **Using simpler models:** Using a simpler model has fewer degrees of freedom and therefore the estimates will be more robust. For example, using logistic regression rather than neural networks.
- **Ensemble approach:** Using one complex model will be more cumbersome thus it is better to use the simpler model and further integrate it with bagging or boosting.
- **No cross-validation data:** This generalizes the idea of using a simple model for a small dataset. If none of the data is small do not use cross-validation of data for hyperparameter optimization.

In addition to the above solution, there are some technical solutions to the small dataset problem which are data augmentation and transfer learning. These two techniques are described in detail below.

5.5.2 Data augmentation

Data augmentation is a technique through which artificial data can be generated with a simple code and can be used to train the model more efficiently. This technique increases the generalizability and convergence of neural networks [96]. In reality, we can make an image dataset that is taken over a realistic condition. But the environment position and contrast can change with the scenario. This change can be tackled more efficiently by the data augmentation method and can increase the dataset library for a better model. The augmentation technique has been grouped into two general categories. One is a traditional approach [97] which is based on the basic methods such as cropping, flip, translation, shearing, contrast change, etc., the second is generative adversarial networks (GANs) [98]. This has the potential to generate a fake image [99]. Thus the output and accuracy can be increased by using both real and augmented data [100].

5.5.3 Transfer learning

Transfer learning, which is also known as fine-tuning, is used for a specific case when there is little data on the task to be performed but there is abundant data on a related problem. Transfer learning is reusing the pretrained model for a new similar problem. The important point for which it has gained popularity is that it can train a neural network with less data. In the basic image processing technique, the neural network detects the edge in the first layer by the use of the kernel, the specific shape is selected in the middle layer, and a specific task is in the last layer. But in the transfer learning process, the pretrained first and middle layer are kept as is and the changes can only be made in the last layer. Through this approach, training time is reduced and the performance of the neural network is increased. Nine pre-trained models are provided by Keras that can be used for fine-tuning, prediction, feature extraction, and transfer learning. Transfer learning is being used in material science for many applications: multiple property prediction [101], microstructure reconstruction [102], predicting phases [103], and crack initiation site [104]. Transfer learning is gaining popularity in the segment of material science, yet the use of this method is not successful because of a lack of data [105]. This problem is due to a lack of data availability and a platform to share the data.

5.6 IMPLEMENTATION OF MACHINE LEARNING IN MATERIAL SCIENCE

Materials are the very basic building block of any industry. In recent times, various scientists are driven toward AI for the problems associated with the development of material science. Image processing is a very simple method in which the image is divided into a specific number of grids. These grids contain a specific region of the image. These regions are selected with the help of the kernel and then identifying the required point from that region [106]. In this chapter, the implementation of machine learning in material science for various specific applications is discussed.

The use of the image for the analysis of various materials has been investigated by going through various research papers. In a recent publication, the strength and failure mechanism of mechanical properties and failure modes of polyamide 12 (PA12) composite which is 3D printed were studied using image processing or digital image correlation [107]. It is well known and proven that with the help of AI and computer vision a computer can replicate the judgment as a human can visualize, thus these tools are better and can perform at par with a human in virtualization tasks with the help of computer vision [108]. The microstructure is an intrinsic property of every material from which various material properties [109] can be implied. In one of the researches, the 3D printing process is monitored with the help of image processing. Thus, by this method, the real-time problem in the printing can be found and the printing can be stopped for a bad print, hence saving time and money [110]. Recently classification and segmentation have been used to process titanium microstructure [111]. The steel material class has also been categorized with image pre-processing and statistical analysis. By using a random forest (RF) algorithm a better prediction/classification can be done as compared to any other model [112]. Multi-class classification for ten different thermomechanical processing conditions of a U-10Mo alloy has been used to evaluate the classification model performance for different microstructure representations [77]. Unsupervised machine learning (ML) has also been used for the identification and characterization of microstructures in three-dimensional (3D) samples obtained from molecular dynamics simulations, particle tracking data, or experiments [113]. This image-processing algorithm can handle a large volume of image data and has been used for quality control for the invention of new steels [114]. But there is a problem with the image processing technique that uses a two-dimensional image which is not realistic as the material has a three-dimensional structure [115]. In an earlier report, the fracture modes in a composite part (which is made by rapid prototyping) during mechanical testing were analyzed by image processing. The image was taken by microscopy and by doing various pre-processing before feeding it into the model to analyze voids on the surface of the composite [116]. Five different rapid prototyped polymers were analyzed experimentally and theoretically in a low-speed collision [117]. The microstructure characterization and reconstruction (MCR) techniques have few limitations in designing materials because there is a loss of information in microstructure representation or while doing the dimensional reduction. Thus, a deep adversarial learning methodology is used that overcomes the limitations of existing MCR techniques [78]. Rapid prototyping is continuously making its place in the manufacturing industry and involved in various fields due to its potential to make a complex part with ease. The reliability of this printed product has to be checked and improved by choosing a cost-efficient and feasible way. Today the ML algorithm is being used in a range of AM procedures and its final part in the era of Industry 4.0 [118]. The recycled sand and mortar mixture is 3D printed and the plastic shrinkage and cracks development are analyzed with the ratio of sand added [119]. Recently, in a work, a set of 425 silver nanoparticles morphologies has been classified in a beautiful way where it has predicted the correlation with the fermi label energy. The image-based description of nanoparticles and material brings us one step closer to using experimental micrographs as inputs for machine learning [120]. In a recent publication, the material bandgap and spectra are predicted with the help of a coloured optical image which is of size 64 x 64 pixels [28]. In this work a variational autoencoder (VAE) is trained on 180,902 optical absorption spectra and with the help of a high-throughput technique, the optical image is prepared [121]. The technique is quite simple in which a materials structure can be converted into a specific image fingerprint which can be used for image processing and does not require any theoretical pre-assessment of the data. The various pieces of work which have been discussed using machine learning are summarized in Table 5.4.

5.7 DATA ECOSYSTEM

The data ecosystem is a combined platform of enterprise infrastructure and an AI-driven data management system. Plenty of tools are available to address various kinds of material data. Some of the most important data are materials data repositories, high-throughput DFT databases [124], and polymer databases [125]. Various other tools like Citrination [126] and Configurable Data Curation System [127] are there which allow the researchers to create a new database quickly. This tool, data services, and software together gave an impulse to materials research with the help of machine learning [26].

5.7.1 MDF

There is always a need for new data in the research field, but with an ample amount of datasets and a strict need for a data management system [128]. A material data facility is a common link between material data producers and clients to aggregate and discover data from various provenances. It automates and streamlines discovery, access data sharing, and analysis (as shown in Figure 5.8). The collection of data is very easy from MDF in your file. The collected data is generally in openly accessible formats. It supports data collection from various platforms such as Globus endpoints, Google Drive, HTTPS accessible data, etc. Thus to publish any data you have to follow three basic steps: sign up for a free Globus account using existing credentials; collect the data into your preferred file; and follow the instructions in the form to publish your dataset. The datasets can be discovered by using the Python program interface. First, the data is explored through MDF forge facilitates that enable unprecedented exploration of indexed datasets. Then the data is aggregated from MDF by using just a few lines of code. Finally, the indexed dataset is automatically analyzed by using the program.

In addition to MDF, the MGI is also a global platform that has changed the way of researching in material science [129]. The value of the model, as well as the experiment, can be enhanced when the experiment is selected very carefully after prediction [130].

5.7.2 DLHub

A device that is used to connect various computer networks is known as a hub. The Data and Learning Hub for science (DLHub) is a learning

Table 5.4 Various applications of machine learning in material science

Through various dataset / microstructure/ Image	work	Material	Model used	Reference
Failure analysis	Mechanical properties and failure modes of 3D printed fiber-reinforced polyamide 12 (PA12) composite lattice structures were studied	Fiber-reinforced polyamide 12 (PA12) composite	Digital image correlation	[107]
Monitoring 3D printing process	Process monitoring of rapid prototyped parts using AI.	ABS and PLA	SVM	[110]
Processing relationships	The microstructure processing relationship has been done from the limited dataset.	Uranium and molybdenum (underdevelopment as nuclear fuel)	Progressive growing GAN and Pix2Pix GAN	[77]
Adaptive characterization	API Keras is used for the characterization of titanium alloy.	Ti-6Al-4V alloy.	CNN	[111]
Microstructural characterization in 3D samples	Characterization of various microstructure types which include polycrystalline materials grains, porous voids, and structures of soft-matter complex solutions.	Polymer 3D Sample.	Unsupervised ML.	[113]
Pattern recognition	The large volume of image data can be handled in a short time which can be used for quality control. It can help in the analysis of new steel for the quest for new steel.	Steel	Fast Random Forest	[114]
Property prediction	Microstructural analysis with the help of machine learning has been done to predict property-based microstructural analysis, property prediction, and properties-to-microstructure inverse analysis.	Steel	MIPHA and rMIPHA	[115]
Tensile and flexural study	Fracture interface study using image processing of 3D printed carbon fiber composite.	Carbon fiber-reinforced polymer composite.	Image processing	[116]

(Continued)

Table 5.4 (Continued)

Through various dataset / microstructure/ Image	work	Material	Model used	Reference
Degree of hydration of concrete	Image of different magnification has been taken which reduces the effort of training and computational time.	Concrete	Random Forest classifier	[122]
Prediction of Vickers hardness	The hardness of a Co- and Ni-based superalloys have been estimated with the help of a microstructural image.	Co- and Ni-based superalloys	Gaussian process regression (GPR)	[123]
Collision test	Polymers under collision with a rigid rod.	PLA, ABS, and acrylic.	Image Processing	[117]
Plastic shrinkage	Plastic shrinkage and cracking of 3D printed mortar mixed with recycled sand as fine aggregates	Mortar and recycled Sand	Image processing	[119]

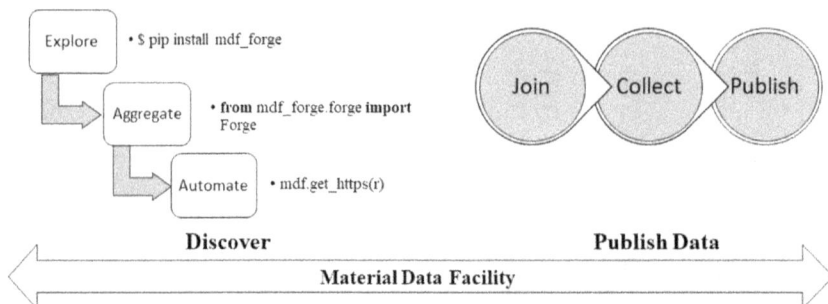

Figure 5.8 Discover and publish data with a material data facility

platform used to tackle the two problems in the ML lifecycle, one is the ML model and the other is associated data. With the help of DLHub, one can publish, share, discover, and reuse ML algorithms. It is a common principle that enables researchers and model creators to share models and at the same time provide credit to the contributor. Thus, DLHub can be used to increase reliability, efficiency, and scalability while integrating it with scientific processes [28]. GitHub is a platform which has multiple modules like TensorFlow [8], Keras [9], and Scikit-learn [10], integrated with Globus [11] which helps in authenticating and inferring from a given dataset [27]. DLHubClient provides a Python wrapper around the DLHub and funcX web services. In this part of the guide, we describe how to use the client to publish, discover, and use servables.

5.8 SWOT ANALYSIS OF ML IN MATERIAL SCIENCE

The strength weakness opportunity threat analysis of machine learning in material science has been done. Although ML has a lot of strengths, the analysis of any microstructure can be done very easily in a shorter time. With the help of ML, the experimental setup is not required on such a large scale and thus there is less wastage of material. This helps in material development and subsequently helps smart production in the era of Industry 4.0. The most common weak point of ML is the lack of dataset. The dataset is the very basic building block of any ML algorithm. Another problem associated with it is that while analyzing any material the image is taken in 2D form, but in the real-world application, the material has a 3D shape. This dimensional contrast creates problems sometimes. In recent times the various data ecosystems have been interconnected thus data assessment becomes easy. This becomes a great opportunity for our research community to gather datasets. Another opportunity is the use of transfer learning in the field of material science. This method has also not been explored

Strength

- Quick analysis of microstructure.
- Save time in materials development.
- Not required experimental setup
- Less wastage of material

Weakness

- Lack of data.
- Material has a 3D form but most analysis is of 2D image, which is not comparable to the real world.

Image processing in Material Science

- Data ecosystem has increased the interconnection between different research community.
- Transfer learning

Opportunity

- Model breakdown can lead to misleading conclusion.

Threat

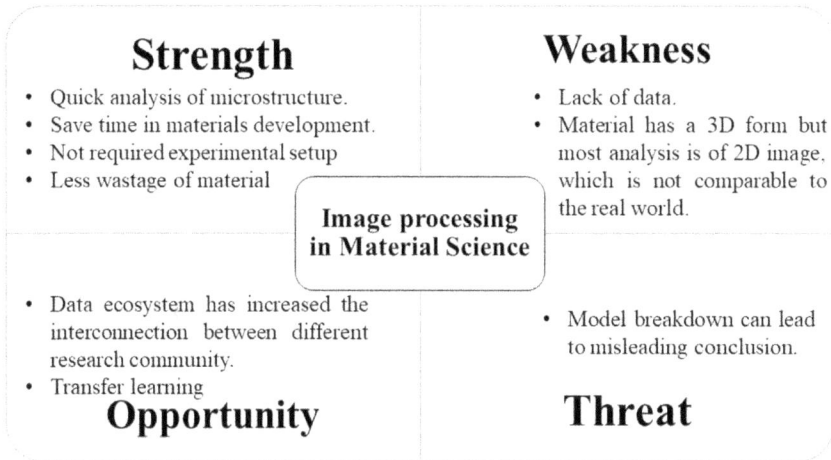

Figure 5.9. SWOT analysis of ML in material science

much yet in this domain. One of the biggest threats is that if any model breaks, then that can lead to a misleading conclusion. This can be dangerous at times if the prediction or analysis of sensitive material is going on. Thus, it is necessary to verify the model and algorithm properly before using it for any sensitive application. The SWOT analysis of ML in material science has been well described in Figure 5.9.

5.9 CONCLUSION AND FUTURE SCOPE

Incorporation of any technology either in research or in the industry has some new challenges. Thus, whether it is machine learning for materials development or Industry 4.0 for better production there is always defiance. But this is how new technology has always evolved. Whenever a new situation arises our brain has been designed such that it can decide from its experience. This experience from the past improvises the decision-making for the present. This is how ML or Industry 4.0 works. This is just a beginning for research and production where both can be automated with less human intervention.

REFERENCES

1. M. Umehara, H. S. Stein, D. Guevarra, P. F. Newhouse, D. A. Boyd, and J. M. Gregoire, "Analyzing machine learning models to accelerate generation of fundamental materials insights," *npj Comput. Mater.*, 5(1), 2019, doi: 10.1038/s41524-019-0172-5.

2. Y. Liu, T. Zhao, W. Ju, and S. Shi, "Materials discovery and design using machine learning," *J. Mater.*, 3(3), pp. 159–177, Sep. 2017, doi: 10.1016/j.jmat.2017.08.002.

3. J. Kim, D. Kang, S. Kim, and H. W. Jang, "Catalyze materials science with machine learning," *ACS Mater. Lett.*, 3(8), pp. 1151–1171, Aug. 2021, doi: 10.1021/acsmaterialslett.1c00204.

4. Z. Zhao and H. Liu, "Spectral feature selection for supervised and unsupervised learning," *ACM Int. Conf. Proceeding Ser.*, 227, pp. 1151–1157, 2007, doi: 10.1145/1273496.1273641.

5. X. Xu, Q. Wei, H. Li, Y. Wang, Y. Chen, and Y. Jiang, "Recognition of polymer configurations by unsupervised learning," *Phys. Rev. E*, 99(4), p. 43307, 2019, doi: 10.1103/PhysRevE.99.043307.

6. "Opportunities challenges of ML in material science.pdf."

7. A. Jain *et al.*, "Commentary: The materials project: A materials genome approach to accelerating materials innovation," *APL Materials*, p. 011002, 2013, doi: 10.1063/1.4812323.

8. T. Acharya, "Image Processing."

9. I. F. P. Energies, M. Paristech, M. Math, P. Ligm, and E. P. France, "Image processing for materials characterization: Issues, challenges and opportunities," doi: 10.1109/ICIP.2014.7025985.

10. R. I. Systems, D. Sarkar, R. Bali, and T. Sharma, "Practical machine learning with Python."

11. S. Guido, "Introduction to machine learning with Python."

12. B. Kailkhura, B. Gallagher, S. Kim, A. Hiszpanski, and T. Y. J. Han, "Reliable and explainable machine-learning methods for accelerated material discovery," *npj Comput. Mater.*, 5(1), pp. 1–9, 2019, doi: 10.1038/s41524-019-0248-2.

13. A. Talapatra, B. P. Uberuaga, C. R. Stanek, and G. Pilania, "A machine learning approach for the prediction of formability and thermodynamic stability of single and double perovskite oxides," *Chem. Mater.*, 33(3), pp. 845–858, 2021, doi: 10.1021/acs.chemmater.0c03402.

14. A. M. Deml, R. O. Hayre, C. Wolverton, and V. Stevanovi, "Predicting density functional theory total energies and enthalpies of formation of metal-nonmetal compounds by linear regression," 085142, pp. 1–9, 2016, doi: 10.1103/PhysRevB.93.085142.

15. F. A. Faber, A. Lindmaa, O. A. Von Lilienfeld, and R. Armiento, "Machine learning energies of 2 million Elpasolite ABC_2D_6 crystals," 135502(Sept.), pp. 2–7, 2016, doi: 10.1103/PhysRevLett.117.135502.

16. A. C. Rajan, A. Mishra, S. Satsangi, R. Vaish, and H. Mizuseki, "Machine-learning-assisted accurate band gap predictions of functionalized MXene," 2018, doi: 10.1021/acs.chemmater.8b00686.

17. Y. Zhuo, A. M. Tehrani, and J. Brgoch, "Predicting the band gaps of inorganic solids by machine learning," 2018, doi: 10.1021/acs.jpclett.8b00124.

18. G. Pilania, C. N. Iverson, T. Lookman, and B. L. Marrone, "Machine-learning-based predictive modeling of glass transition temperatures: A case of polyhydroxyalkanoate homopolymers and copolymers," 2019, doi: 10.1021/acs.jcim.9b00807.

19. M. De Jong, W. Chen, R. Notestine, K. Persson, and G. Ceder, "A statistical learning framework for materials science: Application to Elastic Moduli of k-nary inorganic polycrystalline compounds," *Nat. Publ. Gr.*, (Jun.), pp. 1–11, 2016, doi: 10.1038/srep34256.

20. A. P. Garland, B. C. White, B. H. Jared, M. Heiden, E. Donahue, and B. L. Boyce, "Deep convolutional neural networks as a rapid screening tool for complex additively manufactured structures," *Addit. Manuf.*, 35(Febr.), p. 101217, 2020, doi: 10.1016/j.addma.2020.101217.

21. A. Seko, H. Hayashi, K. Tsuda, L. Chaput, and I. Tanaka, "Prediction of low-thermal-conductivity compounds with first-principles anharmonic lattice-dynamics calculations and Bayesian optimization," 205901(Nov.), pp. 1–5, 2015, doi: 10.1103/PhysRevLett.115.205901.

22. H. Wei, S. Zhao, Q. Rong, and H. Bao, "Predicting the effective thermal conductivities of composite materials and porous media by machine learning methods," *Int. J. Heat Mass Transf.*, 127, pp. 908–916, 2018, doi: 10.1016/j.ijheatmasstransfer.2018.08.082.

23. A. Chowdhury, E. Kautz, B. Yener, and D. Lewis, "Image driven machine learning methods for microstructure recognition," *Comput. Mater. Sci.*, 123, pp. 176–187, 2016, doi: 10.1016/j.commatsci.2016.05.034.

24. G. Pilania, "Machine learning in materials science: From explainable predictions to autonomous design," *Comput. Mater. Sci.*, 193(Mar.), p. 110360, 2021, doi: 10.1016/j.commatsci.2021.110360.

25. K. Tsutsui, H. Terasaki, K. Uto, T. Maemura, and S. Hiramatsu, "A methodology of steel microstructure recognition using SEM images by machine learning based on textural analysis," *Mater. Today Commun.*, 25(Jul.), p. 101514, 2020, doi: 10.1016/j.mtcomm.2020.101514.

26. B. Blaiszik *et al.*, "A data ecosystem to support machine learning in materials science," *MRS Commun.*, 9(4), pp. 1125–1133, 2019, doi: 10.1557/mrc.2019.118.

27. R. Chard *et al.*, "DLHub: Model and data serving for science," *IEEE Int. Parallel Distrib. Process. Symp.*, 2019, pp. 283–292, 2019, doi: 10.1109/IPDPS.2019.00038.

28. R. Chard *et al.*, "Publishing and Serving Machine Learning Models with DLHub," doi: 10.1145/3332186.3332246.

29. T. Le, "Prediction of tensile strength of polymer carbon nanotube composites using practical machine learning method," 2020, doi: 10.1177/0021998320953540.

30. S. Van Der Walt *et al.*, "Scikit-image: Image processing in python," *PeerJ*, 2014(1), pp. 1–18, 2014, doi: 10.7717/peerj.453.

31. B. Evaluation *et al.*, "Using physical parameters for phase prediction of multicomponent alloys by the help of tensorflow machine learning with limited data," *Sak. Univ. J. Sci.*, 25(1), pp. 200–211, 2021.

32. D. Stout, R. Passingham, C. Frith, J. Apel, and T. Chaminade, "Technology, expertise and social cognition in human evolution," *Eur. J. Neurosci.*, 33(7), pp. 1328–1338, Apr. 2011, doi: 10.1111/j.1460-9568.2011.07619.x.

33. I. S. Candanedo, E. H. Nieves, S. R. Gonz.lez, M. T. S. Mart.n, and A. G. Briones, "Machine learning predictive model for industry 4.0," pp. 501–510, 2018.

34. R. Rai, M. K. Tiwari, D. Ivanov, and A. Dolgui, "Machine learning in manufacturing and industry 4.0 applications," *Int. J. Prod. Res.*, 59(16), pp. 4773–4778, 2021, doi: 10.1080/00207543.2021.1956675.

35. F. Ansari, S. Erol, and W. Sihn, "Rethinking human-machine learning in Industry 4.0: How does the paradigm shift treat the role of human learning?" *Procedia Manuf.*, 23(2017), pp. 117–122, 2018, doi: 10.1016/j.promfg.2018.04.003.

36. J. Alzubi, A. Nayyar, and A. Kumar, "Machine learning from theory to algorithms: An overview," *J. Phys. Conf. Ser.*, 1142(1), 2018, doi: 10.1088/1742-6596/1142/1/012012.

37. A. L. Fradkov, "Early history of machine learning," *IFAC PapersOnLine*, 53(2), pp. 1385–1390, 2020, doi: 10.1016/j.ifacol.2020.12.1888.

38. F. Rosenblatt, "The Perceptron: A probabilistic model for information storage and organization in the brain," *Psychol. Rev.*, 65(6), pp. 386–408, 1958, doi: 10.1037/h0042519.

39. *Machine Learning Methods for Planning.* Elsevier, 1993.

40. T. J. Sejnowski and C. R. Rosenberg, "Parallel networks that learn to pronounce English text," *Complex Syst.*, 1, pp. 145–168, 1987. Available: http://cs.union.edu/~rieffelj/classes/2011-12/csc320/readings/Sejnowski-speech-1987.pdf.

41. D. E. Rumelhart, G. E. Hinton, and R. J. Williams, "Learning representations by back-propagating errors," *Nature*, 323(6088), pp. 533–536, Oct. 1986, doi: 10.1038/323533a0.

42. J. Han, Y. Cai, and N. Cercone, "Data-driven discovery of quantitative rules in relational databases," *IEEE Trans. Knowl. Data Eng.*, 5(1), pp. 29–40, 1993, doi: 10.1109/69.204089.

43. M. Abadi *et al.*, "TensorFlow: A system for large-scale machine learning this paper is included in the proceedings of the TensorFlow: A system for large-scale machine learning," 2016.

44. Y. Taigman, M. A. Ranzato, T. Aviv, and M. Park, "Taigman_DeepFace_Closing_the_2014_CVPR_paper," doi: 10.1109/CVPR.2014.220.

45. M. Li, "Scaling distributed machine learning with the parameter server," in *Proceedings of the 11th USENIX Symposium on Operating Systems Design and Implementation*, pp. 583–598, OSDI 2014.

46. J. X. Chen, "The evolution of computing: AlphaGo," *Comput. Sci. Eng.*, 18(4), pp. 4–7, Jul. 2016, doi: 10.1109/MCSE.2016.74.

47. D. Golovin, B. Solnik, S. Moitra, G. Kochanski, J. Karro, and D. Sculley, "Google vizier," in *Proceedings of the 23rd ACM SIGKDD International Conference on Knowledge Discovery and Data Mining*, pp. 1487–1495, Aug. 2017, doi: 10.1145/3097983.3098043.

48. R. Jacobs *et al.*, "The materials simulation toolkit for machine learning (MAST-ML): An automated open source toolkit to accelerate data-driven materials research," *Comput. Mater. Sci.*, 176, p. 109544, Apr. 2020, doi: 10.1016/j.commatsci.2020.109544.

49. B. Mahesh, "Machine learning algorithms - A review," *Int. J. Sci. Res.*, 9(1), 2020, doi: 10.21275/ART20203995.

50. R. Ramprasad, R. Batra, G. Pilania, A. Mannodi-kanakkithodi, and C. Kim, "Machine learning in materials informatics: recent applications and prospects," *npj Comput. Mater.*, (Jul.), 2017, doi: 10.1038/s41524-017-0056-5.

51. F. B. Marin, C. Gurau, and M. Marin, "Machine learning technique to detect defects on the steel surface," 4829, pp. 39–44.

52. P. Valdiviezo-Diaz, F. Ortega, E. Cobos, and R. Lara-Cabrera, "A collaborative filtering approach based on nave Bayes classifier," *IEEE Access*, 7, pp. 108581–108592, 2019, doi: 10.1109/ACCESS.2019.2933048.

53. S. B. Kotsiantis, "Decision trees: A recent overview," *Artif. Intell. Rev.*, 39(4), pp. 261–283, 2013, doi: 10.1007/s10462-011-9272-4.

54. C. Vens, J. Struyf, L. Schietgat, S. D.eroski, and H. Blockeel, "Decision trees for hierarchical multi-label classification," *Mach. Learn.*, 73(2), pp. 185–214, 2008, doi: 10.1007/s10994-008-5077-3.

55. S. Albawi, T. A. Mohammed, and S. Al-Zawi, "Understanding of a convolutional neural network," *Proceedings of the 2017 International Conference on Engineering and Technology*, 2018, pp. 1–6, 2017, doi: 10.1109/ICEngTechnol.2017.8308186.

56. L. Yu, B. Li, and B. Jiao, "Research and implementation of CNN based on TensorFlow," *IOP Conf. Ser. Mater. Sci. Eng.*, 490(4), 2019, doi: 10.1088/1757-899X/490/4/042022.

57. X. Lin, C. Zhao, and W. Pan, "Towards accurate binary convolutional neural network," *Adv. Neural Inf. Process. Syst.*, 3, pp. 345–353, 2017.

58. J. Jang et al., "Residual neural network-based fully convolutional network for microstructure segmentation," *Sci. Technol. Weld. Join*, 25(4), pp. 282–289, 2020, doi: 10.1080/13621718.2019.1687635.

59. J. R. Koza and R. Poli, "Genetic programming." In *Search Methodologies.* Boston, MA: Springer US, pp. 127–164.

60. K. S. Tai, R. Socher, and C. D. Manning, "Improved semantic representations from tree-structured long short-term memory networks," In *Proceedings of the 53rd Annual Meeting of the Association for Computational Linguistics and the 7th International Joint Conference on Natural Language*, 1, pp. 1556–1566, doi: 10.3115/v1/p15-1150.

61. M. Mishra and M. Srivastava, "A view of Artificial Neural Network," In *2014 International Conference on Advances in Engineering & Technology Research (ICAETR - 2014)*, Aug. 2014, pp. 1–3, doi: 10.1109/ICAETR.2014.7012785.

62. J. Kennedy and R. C. Eberhart, "A discrete binary version of the particle swarm algorithm." In *1997 IEEE International Conference on Systems, Man, and Cybernetics. Computational Cybernetics and Simulation*, vol. 5, pp. 4104–4108, doi: 10.1109/ICSMC.1997.637339.

63. I. Okfalisa, M. Gazalba, N. G. I. Reza, "Comparative analysis of k-nearest neighbor and modified k-nearest neighbor algorithm for data classification." In *2nd International Conferences on Information Technology, Information Systems and Electrical Engineering (ICITISEE)*, Nov. 2017, pp. 294–298, doi: 10.1109/ICITISEE.2017.8285514.

64. Z. Majdisova and V. Skala, "Radial basis function approximations: Comparison and applications," *Appl. Math. Modell.*, 51, pp. 728–743, Nov. 2017, doi: 10.1016/j.apm.2017.07.033.

65. D. Chicco, "Siamese neural networks: An overview," 2021, pp. 73–94.

66. W. S. Noble, "What is a support vector machine?," *Nat. Biotechnol.*, 24(12), pp. 1565–1567, Dec. 2006, doi: 10.1038/nbt1206-1565.

67. S. Suthaharan, "Support vector machine," 2016, pp. 207–235.

68. K. P. Sinaga and M.-S. Yang, "Unsupervised K-means clustering algorithm," *IEEE Access*, 8, pp. 80716–80727, 2020, doi: 10.1109/ACCESS.2020.2988796.

69. M. Z. Hossain, M. N. Akhtar, R. B. Ahmad, and M. Rahman, "A dynamic K-means clustering for data mining," *Indones. J. Electr. Eng. Comput. Sci*, 13(2), p. 521, Feb. 2019, doi: 10.11591/ijeecs.v13.i2.pp521-526.

70. G. Douzas and F. Bacao, "Self-organizing map oversampling (SOMO) for imbalanced data set learning," *Expert Syst. Appl.*, 82, pp. 40–52, Oct. 2017, doi: 10.1016/j.eswa.2017.03.073.

71. M. Khanzadeh, P. Rao, R. Jafari-Marandi, B. K. Smith, M. A. Tschopp, and L. Bian, "Quantifying geometric accuracy with unsupervised machine learning: Using self-organizing map on fused filament fabrication additive manufacturing parts," *J. Manuf. Sci. Eng. Trans. ASME*, 140(3), 2018, doi: 10.1115/1.4038598.

72. Y. Nomura, A. S. Darmawan, Y. Yamaji, and M. Imada, "Restricted Boltzmann machine learning for solving strongly correlated quantum systems," *Phys. Rev. B*, 96(20), p. 205152, Nov. 2017, doi: 10.1103/PhysRevB.96.205152.

73. J. E. van Engelen and H. H. Hoos, "A survey on semi-supervised learning," *Mach. Learn.*, 109(2), pp. 373–440, 2020, doi: 10.1007/s10994-019-05855-6.

74. S. R. Ahmed Ahmed, I. Al Barazanchi, Z. A. Jaaz, and H. R. Abdulshaheed, "Clustering algorithms subjected to K-mean and gaussian mixture model on multidimensional data set," *Period. Eng. Nat. Sci.*, 7(2), p. 448, Jun. 2019, doi: 10.21533/pen.v7i2.484.

75. V. Fran,ois-Lavet, P. Henderson, R. Islam, M. G. Bellemare, and J. Pineau, "An introduction to deep reinforcement learning," *Found. Trends® Mach. Learn.*, 11(3–4), pp. 219–354, 2018, doi: 10.1561/2200000071.

76. Liu, X.; Song, L.; Liu, S.; Zhang, Y. A Review of Deep-Learning-Based Medical Image Segmentation Methods. Sustainability 2021, 13, 1224. https://doi.org/10.3390/su13031224.

77. W. Ma *et al.*, "Image-driven discriminative and generative machine learning algorithms for establishing microstructure-processing relationships," *J. Appl. Phys.*, 128(13), 2020, doi: 10.1063/5.0013720.

78. X. Li, Z. Yang, L. C. Brinson, A. Choudhary, A. Agrawal, and W. Chen, "A deep adversarial learning methodology for designing microstructural material systems," 2018, pp. 1–14, doi: 10.1115/detc2018-85633.

79. B. Jang, M. Kim, G. Harerimana, and J. W. Kim, "Q-learning algorithms: A comprehensive classification and applications," *IEEE Access*, 7, pp. 133653–133667, 2019, doi: 10.1109/ACCESS.2019.2941229.

80. Z.-H. Zhou, "Ensemble learning." In *Machine Learning*. Singapore: Springer, pp. 181–210, 2021.

81. F. Stulp and O. Sigaud, "Many regression algorithms, one Unified Model: A review," *Neural Netw.*, 69, pp. 60–79, Sep. 2015, doi: 10.1016/j.neunet.2015.05.005.

82. D. W. Aha, D. Kibler, and M. K. Albert, "Instance-based learning algorithms," *Mach. Learn.*, 6(1), pp. 37–66, Jan. 1991, doi: 10.1007/BF00153759.

83. F. Girosi, M. Jones, and T. Poggio, "Regularization theory and neural networks architectures," *Neural Comput.*, 7(2), pp. 219–269, Mar. 1995, doi: 10.1162/neco.1995.7.2.219.

84. Xie Niuniu and Liu Yuxun, "Notice of retraction: Review of decision trees," In *2010 3rd International Conference on Computer Science and Information Technology*, pp. 105–109, Jul. 2010, doi: 10.1109/ICCSIT.2010.5564437.

85. M. E. Tipping, "Bayesian inference: An introduction to principles and practice in machine learning," In: Bousquet, O., von Luxburg, U., R.tsch, G. (Eds.) Advanced Lectures on Machine Learning. ML 2003. *Lecture Notes in Computer Science*, vol 3176. Springer, Berlin, Heidelberg. https://doi.org/10 .1007/978-3-540-28650-9_3.

86. C. F. Eick, N. Zeidat, and Z. Zhao, "Supervised clustering - Algorithms and benefits." In *16th IEEE International Conference on Tools with Artificial Intelligence*, pp. 774–776, doi: 10.1109/ICTAI.2004.111.

87. M. Shahin *et al.*, "Big data analytics in association rule mining: A systematic literature review," In *2021 the 3rd International Conference on Big Data Engineering and Technology (BDET)*, pp. 40–49, 2021, doi: 10.1145/3474944.3474951.

88. M. Dixit, A. Tiwari, H. Pathak, and R. Astya, "An overview of deep learning architectures, libraries and its applications areas," *2018 International Conference on Advances in Computing, Communication Control and Networking (ICACCCN)*, pp. 293–297, Oct. 2018, doi: 10.1109/ ICACCCN.2018.8748442.

89. G. T. Reddy *et al.*, "Analysis of dimensionality reduction techniques on big data," *IEEE Access*, 8, pp. 54776–54788, 2020, doi: 10.1109/ ACCESS.2020.2980942.

90. "Machine learning with Python." www.tutorialspoint.com (accessed Dec. 27, 2021).

91. A. Choudhury, "The role of machine learning algorithms in materials science: A state of art review on Industry 4 . 0," *Arch. Comput. Methods Eng.*, 89(01234567), 2020, doi: 10.1007/s11831-020-09503-4.

92. S. S. Ribeiro, A. Ferrasa, and R. Falate, "'Using Python with Simplecv to detect a corn kernel in digital using Python with Simplecv to detect A,' no. Aug., 2014. V.4, N.2, Aug/2014, ISSN 2237-4523

93. S. Raschka, J. Patterson, and C. Nolet, "Machine learning in python: Main developments and technology trends in data science, machine learning, and artificial intelligence," *Information*, 11(4), p. 193, Apr. 2020, doi: 10.3390/ info11040193.

94. Y. Zhang and C. Ling, "A strategy to apply machine learning to small datasets in materials science," *npj Comput. Mater.*, 4(1), pp. 28–33, 2018, doi: 10.1038/s41524-018-0081-z.

95. S. K. Kauwe, J. Graser, R. Murdock, and T. D. Sparks, "Can machine learning find extraordinary materials?" 174(Aug.), 2020, doi: 10.1016/j. commatsci.2019.109498.

96. M. Alber *et al.*, "iNNvestigate neural networks!," *J. Mach. Learn. Res.*, 20, pp. 1–8, 2019.

97. C. Shorten and T. M. Khoshgoftaar, "A survey on image data augmentation for deep learning," *J. Big Data*, 2019, doi: 10.1186/s40537-019-0197-0.

98. Elgendi M., Nasir M.U., Tang Q., Smith D., Grenier J.-P., Batte C., Spieler B., Leslie W.D., Menon C., Fletcher R.R., Howard N, Ward R, Parker W. and Nicolaou S (2021) The Effectiveness of Image Augmentation in Deep Learning Networks for Detecting COVID-19: A Geometric Transformation Perspective. Front. Med. 8:629134. doi: 10.3389/fmed.2021.629134.

99. S. Medghalchi, C. F. Kusche, E. Karimi, U. Kerzel, and S. Korte-kerzel, "Damage analysis in dual-phase steel using deep learning: Transfer from uniaxial to biaxial straining conditions by image data augmentation," *JOM*, 72(12), pp. 4420–4430, 2020, doi: 10.1007/s11837-020-04404-0.

100. B. Ma *et al.*, "Data augmentation in microscopic images for material data mining," *npj Comput. Mater.*, 2020, doi: 10.1038/s41524-020-00392-6.

101. D. Guevarra and C. P. Gomes, "Materials representation and transfer learning for multi-property prediction Materials representation and transfer learning for multi-property prediction," 021409(Febr.), 2021, doi: 10.1063/5.0047066.

102. Li, X., Zhang, Y., Zhao, H. et al. A Transfer Learning Approach for Microstructure Reconstruction and Structure-property Predictions. *Sci Rep* 8, 13461 (2018). https://doi.org/10.1038/s41598-018-31571-7

103. S. Feng, H. Fu, H. Zhou, Y. Wu, Z. Lu, H. Dong, "Open a general and transferable deep learning framework for predicting phase formation in materials," *npj Comput. Mater.*, pp. 1–10, doi: 10.1038/s41524-020-00488-z.

104. S. Y. Wang and T. Guo, "Transfer learning-based algorithms for the detection of fatigue crack initiation sites: A comparative study," 8(Nov.), pp. 1–13, 2021, doi: 10.3389/fmats.2021.756798.

105. H. Yamada *et al.*, "Predicting materials properties with little data using shotgun transfer learning," 2019, doi: 10.1021/acscentsci.9b00804.

106. A. H. Khan, S. S. Sarkar, and R. Sarkar, "A genetic algorithm based feature selection approach for microstructural image classification," May 2021Experimental Techniques 46(9), DOI:10.1007/s40799-021-00470-4.

107. W. Hao, Y. Liu, T. Wang, G. Guo, H. Chen, and D. Fang, "Failure analysis of 3D printed glass fiber/PA12 composite lattice structures using DIC," *Compos. Struct.*, 225(Apr.), p. 111192, 2019, doi: 10.1016/j.compstruct.2019.111192.

108. E. A. Holm *et al.*, "Overview: Computer vision and machine learning for microstructural characterization and analysis," *Metall. Mater. Trans. A Phys. Metall. Mater. Sci.*, 51(12), pp. 5985–5999, 2020, doi: 10.1007/s11661-020-06008-4.

109. M. Karamad, R. Magar, Y. Shi, S. Siahrostami, I. D. Gates, and A. Barati Farimani, "Orbital graph convolutional neural network for material property prediction," *Phys. Rev. Mater.*, 4(9), p. 093801, Sep. 2020, doi: 10.1103/PhysRevMaterials.4.093801.

110. U. Delli and S. Chang, "Automated process monitoring in 3D printing using supervised machine learning," *Procedia Manuf.*, 26, pp. 865–870, 2018, doi: 10.1016/j.promfg.2018.07.111.

111. A. Baskaran, G. Kane, K. Biggs, R. Hull, and D. Lewis, "Adaptive characterization of microstructure dataset using a two stage machine learning approach," *Comput. Mater. Sci.*, 177(Mar.), p. 109593, 2020, doi: 10.1016/j.commatsci.2020.109593.

112. D. S. Bulgarevich, S. Tsukamoto, T. Kasuya, M. Demura, and M. Watanabe, "Automatic steel labeling on certain microstructural constituents with image processing and machine learning tools," *Sci. Technol. Adv. Mater.*, 20(1), pp. 532–542, 2019, doi: 10.1080/14686996.2019.1610668.

113. H. Chan, M. Cherukara, T. D. Loeffler, B. Narayanan, and S. K. R. S. Sankaranarayanan, "Machine learning enabled autonomous microstructural characterization in 3D samples," *npj Comput. Mater.*, 6(1), 2020, doi: 10.1038/s41524-019-0267-z.

114. D. S. Bulgarevich, S. Tsukamoto, T. Kasuya, M. Demura, and M. Watanabe, "Pattern recognition with machine learning on optical microscopy images of typical metallurgical microstructures," *Sci. Rep.*, 8(1), pp. 3–9, 2018, doi: 10.1038/s41598-018-20438-6.

115. Z. L. Wang and Y. Adachi, "Property prediction and properties-to-microstructure inverse analysis of steels by a machine-learning approach," *Mater. Sci. Eng. A*, 744(Dec.), pp. 661–670, 2019, doi: 10.1016/j.msea.2018.12.049.

116. N. Maqsood and M. Rimaauskas, "Tensile and flexural response of 3D printed solid and porous CCFRPC structures and fracture interface study using image processing technique," *J. Mater. Res. Technol.*, 14, pp. 731–742, 2021, doi: 10.1016/j.jmrt.2021.06.095.

117. K. Kardel, H. Ghaednia, A. L. Carrano, and D. B. Marghitu, "Experimental and theoretical modeling of behavior of 3D-printed polymers under collision with a rigid rod," *Addit. Manuf.*, 14, pp. 87–94, 2017, doi: 10.1016/j.addma.2017.01.004.

118. Goh, G.D., Sing, S.L. & Yeong, W.Y. A review on machine learning in 3D printing: applications, potential, and challenges. Artif Intell Rev 54, 63–94 (2021). https://doi.org/10.1007/s10462-020-09876-9

119. H. Zhang and J. Xiao, "Plastic shrinkage and cracking of 3D printed mortar with recycled sand," *Constr. Build. Mater.*, 302(Jul.), p. 124405, 2021, doi: 10.1016/j.conbuildmat.2021.124405.

120. B. Sun and A. S. Barnard, "Texture based image classification for nanoparticle surface characterisation and machine learning," *J. Phys. Mater.*, 1(1), 2018, doi: 10.1088/2515-7639/aad9ef.

121. H. S. Stein, D. Guevarra, P. F. Newhouse, E. Soedarmadji, J. M. Gregoire, "Machine learning of optical properties of materials – Predicting spectra from images and images from spectra," 10(1), 2019, doi: 10.22002/D1.1103.

122. S. S. Bangaru, C. Wang, M. Hassan, H. W. Jeon, and T. Ayiluri, "Estimation of the degree of hydration of concrete through automated machine learning based microstructure analysis – A study on effect of image magnification," *Adv. Eng. Inform.*, 42(Aug.), p. 100975, 2019, doi: 10.1016/j.aei.2019.100975.

123. N. Khatavkar, S. Swetlana, and A. K. Singh, "Accelerated prediction of Vickers hardness of Co- and Ni-based superalloys from microstructure and composition using advanced image processing techniques and machine learning," *Acta Mater.*, 2020, doi: 10.1016/j.actamat.2020.06.042.

124. G. Ceder, H. L. Tuller, "Coarse-graining and data mining approaches to the prediction of structures and their dynamics by Stefano Curtarolo," 1995, 2003, http://dspace.mit.edu/handle/1721.1/7582.

125. A. Mannodi-kanakkithodi *et al.*, "Scoping the polymer genome: A roadmap for rational polymer dielectrics design and beyond," *Mater. Today*, 21(7), pp. 785–796, 2018, doi: 10.1016/j.mattod.2017.11.021.

126. J. O. Mara, B. Meredig, and K. Michel, "Materials data infrastructure: A case study of the citrination platform to examine data import, storage, and access," 68(8), pp. 2031–2034, 2016, doi: 10.1007/s11837-016-1984-0.

127. A. Dima *et al.*, "Informatics infrastructure for the materials genome initiative," 68(8), pp. 2053–2064, 2016, doi: 10.1007/s11837-016-2000-4.

128. B. Blaiszik, K. Chard, J. Pruyne, R. Ananthakrishnan, S. Tuecke, and I. Foster, "The materials data facility: Data services to advance materials science research," 68(8), pp. 2045–2052, 2016, doi: 10.1007/s11837-016-2001-3.

129. J. E. Gubernatis and T. Lookman, "Machine learning in materials design and discovery: Examples from the present," 120301, pp. 1–15, 2018, doi: 10.1103/PhysRevMaterials.2.120301.

130. A. White, "The materials genome initiative: One year on," *MRS Bulletin*, 37(Aug.), pp. 715–716, 2012.

Chapter 6

Blockchain, artificial intelligence, and big data
Advanced technologies for Industry 4.0

Rishabh Machhan, Rupen Trehan, Perminderjit Singh, and Kuldip Singh Sangwan

6.1 INTRODUCTION

The world around us is changing at a rapid pace. New technologies are also coming in no time. Automation is the focal point of the research and development for organizations to achieve. Some countries have managed to achieve automation in various industries. Industry 4.0 deals with the digitalizing of the existing manufacturing processes or the industry practices with the help of advanced intelligent technologies. Prior to Industry 4.0, three industrial revolutions already took place. In the first industrial revolution (I1.0) (late 18th–late 19th century), the transition took place from hand production to machine production with the help of a newly developed steam engine by Sir James Watt. Sir James Watt laid the foundation of mechanical engineering. During this period the concept of the industry came into the picture. Also, this industrial revolution is generally known as the mechanical revolution. In the late 19th century, the second industrial revolution (I2.0) (late 19th–mid 20th century) took place. Due to the discovery of electrical energy, the production engineers got an opportunity to redesign the production line, increasing productivity. During this period, the mass production concept was introduced. After the two world wars, in the 1960s, the first Z1 computer was produced using binary floating points and Boolean logic. The development of this computer laid the foundation for the development of the much-advanced computer that we see today. Due to this, the digital revolution (I3.0) (third industrial revolution) took place in the industry. Computers and electronic devices were integrated into the industry, which helped in monitoring and handling. Implementing Industry 4.0 helped us to use the technologies that help in data storage, management, analysis, and provide a secure network system over which data can be transferred securely.

Now, we are living in the Industry 4.0 era. Industry 4.0 (I4.0) is heavily influenced by advanced technologies. Blockchain provides a platform over which information can be shared with a high level of trust. Information shared will be stored with all the participating members, improving the collaboration among the partners. Blockchain also tracks the transactions

DOI: 10.1201/9781003246466-6

and is immutable, which is one of the main reasons this technology is used for financial transactions. It also provides decentralization, encryption for the ledger of computer filling, to meet the requirement of Industry 4.0, i.e., automation. Artificial intelligence (AI) is one of the critical technologies, which helps in decision-making at a high level of accuracy. Artificial intelligence is an essential technology for Industry 4.0 to be fully functional. Industry 4.0 is strongly dependent on automation and less intervention is required by humans. To achieve that, we need to have good artificial intelligence for training the machines with machine learning. So, it becomes essential for companies to spend their money on bringing and implementing this technology in their industry. For making a better decision, an enormous amount of data is required to be communicated in real-time. For this purpose, big data technology is a critical technology that helps in providing the data in real-time. Big data provides a massive amount of high-velocity data and provides variable data.

However, big data, blockchain, and artificial intelligence have more advantages over other technologies as they can handle large amounts of data and vary the process for customization without making massive changes. These technologies are vital for the adoption of Industry 4.0. In this chapter, we will study how technologies like blockchain, artificial intelligence, and big data help automate the industry and the challenges associated with these technologies while implementing them.

6.2 BLOCKCHAIN

Satoshi Nakamoto introduced the concept of a decentralized blockchain in 2008. However, the existence of Nakamoto is still not confirmed as to whether he is a real or fictional person. He uses the required Hashcash-like method to timestamp the blocks, eliminating trusted third-party verification. Blockchain is a distributed, decentralized, and public ledger consisting of blocks that stores information. The record transaction is stored in many computers, which reduces the alteration in previous transactions. The data is automated, using a peer-to-peer distribution network and timestamp [1]. It consists of several layers: infrastructure, networking, consensus, data, and application.

Blocks are a vital component of this technology that holds valid hashed transactions and is encoded into the Merkle tree. The block consists of the hash address of the previous block that helps to link the chain, and all the members can see this information in the network, which provides transparency. This iterative process verifies the connection of the previous block back to the initial block, known as the genesis block. The members in the network can verify the transaction on their own, which does not cause much money [2]. Decentralization helps the organizations to control activities even at floor level to make better decisions that can improve performance. Transparency helps different organizations to work together and share data in a secure channel and improve their performance. Still, there are few challenges for the adoption of blockchain. Blockchain ensures that each piece of information is shared only once in the network and removes the long-lasting double sharing problem.

6.2.1 Blockchain in industry

Blockchain provides the privacy and security in the autonomous vehicles because of its superiorities like immutability, reliability, transparency, and traceability. Immutability provides the benefits that the information once stored in the block will not be altered by anyone. If someone tries to alter the information in the blockchain then the whole blockchain has to be altered which is impossible to do. As the data is stored in more than one computer, the failure of the network from one node is eliminated. This makes the system more reliable for the application. As the record of the data is unaltered and everyone is able to see the data in the network, it becomes easier to trace the data [3]. Blockchain technology can be used for the transaction purpose among the supply chain partners and provides a reliable and trusted platform where alteration in transaction is not possible. Mehta et al. [4] proposed the use of blockchain in the royalty contract transaction in the oil and gas industry which provides a secure and reliable network. They also compared the results that they received on utilizing the blockchain for the transaction to the existing transaction scheme. They found out that their proposed method is better than the existing method. For achieving business competence, it is essential to have a system where the complete information and knowledge generated throughout the product life is communicated smoothly. Blockchain provides a secure platform for stakeholders to give openness, interoperability decentralization to their business. Liu et al. [5] in their study proposed a blockchain-based customized information service. They also apply this proposed method in co-design, co-creation, proactive maintenance, and regulated recycling between cooperating partners. The results show that the method is scalable and efficient.

Data security or network security is essential for any organization while automating the organization. The existing security framework is not sufficient to address these issues. A blockchain-based system for safe authentication ensures the fine grain security policies. This gives the anonymous auditing, confidentiality, and authentication [6]. Blockchain is also used to secure the IoT (Internet of Things) sensors communication Industry 4.0. Rathi et al. [7] simulated the blockchain secured IoT sensors and compared the results on the basis of chances of attack success, ease of attack identification by the system and delay in authentication. The authentication has been appraised by blockchain.

It is essential to have a clear awareness of the enablers of blockchain for its implementation in industry. The key enablers and drivers for blockchain are cybersecurity, 3D printing, Internet of Things, robotics, cloud computing, and intelligence data analytics [8]. Blockchain with Internet of Things brings the advancements in industries like oil and gas, manufacturing, construction, and engineering, etc. to renovate the existing commercial model to a swifter one [9]. Industry is going through a new industrial revolution i.e., Industry 4.0. This brings major changes in the complete industrial structure. Blockchain technology in this environment is meant to speed up the innovativeness and automation in the industry by improving the concepts associated with blockchain and Industry 4.0. In the Industry 4.0 environment there is a huge amount of data sharing, collection, and analyzing which helps in bringing out the useful information. Pinheiro et al. [10] proposed the multi-agent systems and blockchain to assist in representing an entity in a network of discrete entities and provide the valuable knowledge which will be supportive for making appropriate decisions. They first introduced the model with the aim of resolving the challenges faced by the previously described model. The proposed model is faster, and it improves communication among entities by accessing the needs of the other entity in the network. This helps in providing needful information to the entity with low latency and highly qualitative data. Blockchain finds its application in business processing management, with the target to enhance the processes to get improved performance from the system, such as quick response, better services, and higher profit. In an Industry 4.0 environment, business processing management systems need to have automatic and digitalized business process workflows and support that provide the interoperability transparency of the service sellers. The crucial part of advanced business processing systems is evaluating, transforming, and verifying the reliability and digital assets. Viriyasitavat et al. [11] investigate the automated business processing management (BPM) solutions for selection and composing services in the open business. They use the blockchain technology in the BPM framework and explore how blockchain can be joined to enable reliable, prompt, lucrative analysis and transmission of excellence of services in the workflow composition in management. Agriculture is a crucial sector in the case of social-economic stability. Having transparency and food safety in the food supply chain is still a meaningful concern. The Internet of Things and blockchain is gaining popularity due to their versatility. We can also use blockchain technology in the agricultural sector. A huge amount of data is gathered from various sources that can be optimized and utilized with the help of deep learning methods. For transparency in supply chain management, blockchain is an essential technology that can be used for processes like broadening visibility, digitalization, provenances, smart contracts, and disintegration. Khan et al. [12] took the blockchain and Internet of Things data of Industry 4.0 in the agricultural sector. With the help of advanced deep learning methods, they proposed the hybrid model based on recurrent neural networks. They use the long short-term memory (LSTM) and gated recurrent units (GRU) as the forecasting model and genetic algorithm, which will be used to optimize the parameters of hybrid models. The selection of training parameters is made with the help of a genetic algorithm, and then the long short-term memory is cascaded with gated recurrent units. This proposed model provides consumers the facility to validate their food before they consume it. They can also track the sources and supply chain of the food from remote locations. With the advantages associated with blockchain, IoT, and deep learning techniques, the user can improve their business by predicting the upcoming trends and food demands based on the previous year's data. They can also control the assets in the shown immutable private record. The information stored in the ledger is encrypted, and the Internet of Things enables supply trucks, retail shops, and warehouses, which helps ensure that the product is stored at the appropriate temperature.

6.2.2 Challenges for blockchain implementation

There are many challenges that organizations face while implementing the blockchain in the industries like autonomous vehicles, oil and gas, supply chain, securities, etc. Major challenges that organizations face are as follows:

6.2.2.1 System throughput

Blockchain algorithm requires the high computational power to synchronize and update data while sharing with others. In the case of movable or small nodes computation on site it is not possible and the data latency may occur in the networking system [3].

6.2.2.2 Lack of awareness

For utilizing any technology or method it is essential to know how that technology works. Currently, there are very few people that know this technology. This lack of awareness hinders the adoption process. The investors will not invest their money for bringing the technology in the system because of lack of awareness and understanding. This overall affects the Industry 4.0 adoption process [9].

6.2.2.3 Scalability

Blockchain is a decentralized method where the security of the network is not the responsibility of any central party. Instead, each node is responsible for the network security and they have to update each and every transaction that happens in the network. This requires relatively large storage and increases the size of the block. This also brings the security issue with it as the complete history of data is stored at each node [9, 3].

6.2.2.4 Lack of governance and standards

Due to decentralization of the blockchain, there is no one who is responsible for maintaining the standards in the network. In the court of law, there is no law or article that talks about the standards and norms for blockchain like technologies [9].

6.2.2.5 Authentication

In the case of blockchain for autonomous vehicles, the blockchain network used is a public blockchain where any autonomous vehicle can do the proof of the work and join the network. Due to the openness of the blockchain the possibility of malicious vehicles or nodes added to the network increases. To minimize this, the authentication process has to be completed in real-time otherwise it brings latency in the data interaction and may cause fatal issues in the networks like autonomous vehicles which share data for smooth operation and avoid collision [3].

6.2.2.6 Energy consumption

Energy consumption for the verification of the transaction in the blockchain network is high. As this has been seen that the mining of the coins in cryptocurrency requires high energy to solve the complex problem. All cryptocurrency works on blockchain. High computational power is required to solve complex mathematical problems which eventually requires a large amount of energy [9].

6.2.2.7 Real-time implementation

To provide the information in real-time to the blockchain network, the authentication process has to be done as fast as possible. It also brings a few challenges with it like 51 percent attack. In cases like 51 percent attack, a group of people or nodes get the control of the blockchain network's hash power by more than 50 percent, which makes them powerful enough to manipulate the network as per their benefits [13].

6.2.2.8 Implementation cost

The organizations have to spend large amounts of money to bring the technology within the organization. Before bringing the technology, infrastructure needs to be updated to meet the requirement. Organizations have to train their current employees, so that they can implement the technology efficiently [13].

6.2.2.9 *Lack of knowledge and infrastructure*

The software developers who use this technology are few which affect the implementation of the technology as not all organizations have the privilege to have employees that can utilize technology like blockchain. Also, many organizations lack the infrastructure that is required for the adaptation of the technology. Current IoT-based protocol requires centralized management which makes blockchain adoption difficult [13].

6.3 ARTIFICIAL INTELLIGENCE

Artificial intelligence (AI) is the cognitive intelligence shown by machines similar to the intelligence displayed by humans and animals. It helps in enabling the machines to perform tasks controlled by computers. AI scans and monitors its environment and performs the action that helps accomplish goals. The function of artificial intelligence is inspired by the human mind's learning and problem-solving ability. The development of AI started with finding step-by-step reasoning that human beings use while solving puzzles. For uncertain and half information problem case scenarios, the researcher uses the concept of probability [14]. Organizations are looking to integrate machines with artificial intelligence, which helps machines make decisions as per the data given. Artificial intelligence enhances the decision-making ability as it processes and analyses an enormous amount of data in real-time. AI can perform tedious and time-consuming tasks in real-time and help organizations make decisions at a much faster rate. This technology makes a more straightforward implementation of Industry 4.0 in industries.

Machine learning (ML) is a fundamental concept of AI research. In ML, we use supervised and unsupervised learning methods. In unsupervised learning, we find the pattern in input data, whereas, in the case of supervised learning, we had to label the input data. The variation in supervised learning is of two types, i.e., classification and numerical regression. Where classification means what class or category something belongs to, and in regression, we try to develop the relation between input and output [15]. Artificial intelligence uses natural language processing (NLP), which allows the machines to read and recognize the human language [16]. Autonomous robots are heavily dependent on artificial intelligence. Modern-day robots perform their work or task with minimal human intervention [17].

6.3.1 Artificial intelligence in industry

Recently, the industry has started to acknowledge AI and ML (machine learning), which can help in improving the process's effectiveness. Industrial AI is a systematic method, developing, validating, and deploying the machine learning algorithm to improve machines' performance sustainably. The key components of industrial artificial intelligence are distinguished by ABCDE. These components include analytics technology (A), big data (B), cyber or cloud technology (C), domain know-how (D), and evidence (E) [18]. Artificial intelligence's application can be seen in material selection applied to manufacturing. Supervised learning requires more computation demand, and more complex results can be obtained. AI helps in extracting the data from a massive database. It also helps in learning non-linear functions [19]. Kebisek et al. [20] proposed the elucidation of the artificial intelligence (AI) platform to validate the appropriateness of neural networking to forecast the quality of paint from the selected factors.

Predicting the maintenance in industries is very helpful for businesses to grow in the modern era – intelligence-based preservation scheduling for parallel multi-component over machines during manufacturing. O'Donovan et al. [21], in their work, monitor the performance and reliability of machines in real-time. He later concludes that the reliability and latency cyber-physical system interface designed with centralized and decentralized computation technologies to facilitate real-time monitoring reduces the communication failure up to 6.6 percent under various communication load conditions. Artificial intelligence can optimize the accuracy and rate of packaging and lower the cost of the solution for packaging robots. Chen et al. [22] suggested a particle swarm optimization (PSO) framework for less time-consuming trace planning of robots in intelligence packaging systems. AI can also be used for managing human resources. This will provide a sustainable competitive advantage and improve the employees' skills and management strategy by learning from past cases. AI will help track the performance, efficiency, and pattern of work for both humans and machines and automatically update that information. Results from the data collected will help the organization improve its strategy and increase the productivity of humans and machines [23]. The AI framework helps predict the product's quality throughout its lifespan in the intelligence industry, which eventually improves material utilization comprehensively. With the assistance of the AI model, we can provide the required quality of the product, and based on the previous data, we can predict the future quality requirement via enhancing the existing forecasting method. In Industry 4.0, the management of quality data collected is subcategorized as Quality 4.0 [24]. As the data is collected from different resources, the abundance of quality data is there, which helps the progressive analytics become extremely feasible to forecast the quality and product yield. We also required data communication in real-time throughout the product and ensured quality, product acquiesce, and customer value. The integration of AI and robotics provide a powerful combination for automatic tasks.

In the smart industries, the integration of AI and robotics have arisen as an influential collaboration for digitalizing the industry. Robotics plays an essential role in Industry 4.0, which brings the new paradigm in the industry which is beyond its capabilities and possible with the integration with artificial intelligence (AI) [25]. Different organizations are working in developing the robots to replace human involvement and increase the level of autonomy within the organization. We can use the robots in the production processes in the intelligent industries, which brings the maximum wastage of power with the help of data collected and results from the artificial intelligence algorithm [26]. Organizations are developing robots to improve the effectiveness of assembly and manufacturing units, packaging, and removing the production task scheduling. AI has improved the effectiveness of the robots that have already been proven. In AI, if we integrate new technologies like vision systems, this will enable the robots with more flexibility, reducing the efforts to learn, define, and optimize the complex network in the manufacturing sectors, such as aerospace [27]. The application of robots in the industry is first seen in the packaging sector, where the robots pack products. If we combine the AI with the existing robots or advanced robots for packaging, we can expect to improve the quality, accuracy, packaging speed, and lower the cost of production [28]. Earlier, we used a fixed number of robots and tools in the assembly lines in industries. We can recover and re-create the assembly line with the help of re-configurable tools and movable robots programmed with intrinsic work scheduling mechanisms and smart controlled units with the help of AI. Parente et al. [29] conduct the survey for analyzing the barriers of Industry 4.0 towards scheduling assignments in industries and providing suitable solutions for human-robot integration, which can provide the relevant results for decision making.

6.3.2 Challenges for artificial intelligence implementation

For the implementation of artificial intelligence, organizations have to face numerous challenges. A few of these challenges are discussed below:

6.3.2.1 Data availability

Data is the nucleus of an AI system. Cloud platforms and super-computing extract the information from the raw data in real-time. Artificial intelligence is valid and gets meaningful insights from the data given. For that to work in proper and in an efficient way, sufficient data quality is required [27]. As the data is collected from various resources the data have heterogeneous quality. To get the information out of these data, sometimes intensive inter-connectivity with computational platforms [29]. Data available to train and validate the model is still not sufficient. Technologies like deep learning and machine learning require massive data for training the model to give unfiltered results [30].

6.3.2.2 Cybersecurity

With the implementation of advanced technologies in the industry, we connect most systems to the internet or cyber systems. So, it makes systems vulnerable to cyberattacks that can damage the whole system, and not much has been done significantly to minimize or counter this threat [31]. The high-level overview of common adversarial attacks can affect organizations by corrupting the data and robbing the models [27]. There is also a risk of reverse engineering attacks that can obtain the knowledge and sensitive data right from the model [30].

Large organizations require a security strategy and risk management system that are suitable for their specific security demands. The purpose of this article is to discuss the various challenges that businesses should deal with in order to protect themselves against cyber-attacks. Production has been more digitalized, paving the door for approaches such as the IoT, augmented reality, and many more to emerge.

6.3.2.3 M2M interactions

AI algorithms can map the set of input data to output data accurately. Still, these algorithms are also susceptible to slight variations in the dataset that can be caused by deviations from machine to machine. It becomes essential to ensure that individual AI will not interfere or get engaged with another AI in the network [18].

6.3.2.4 Transfer learning

Humans can implement the knowledge they have acquired in the previous task to solve new problems, which helps solve the problem more easily or provide better solutions. A similar kind of knowledge transfer is required in AI, which helps find the results from even the scarce data. This transfer learning is split into three: working situations, various machines/stations, and various types of machine failure [30].

6.3.3.5 Data quality

For training the model, the dataset required should be clean with a minimum bias factor. If the model is trained on biased and unfiltered data, then the results that we get will have more flaws [18]. The more superb the quality of the data, the more accurate the AI algorithm will show. Not having access to high-quality data sets will lead to substandard AI systems. Also, transparency in the product quality can be supervised with the help of AI-based machine learning, which helps detect outliers in the dataset and provide a standard quality [27]. Data quality can be characterized into four main dimensions: intrinsic, contextual, representational, and accessibility. Minimal research has been carried out for assessing these dimensions, which eventually act as a barrier [29].

6.3.2.6 Data accessibility

Data accessibility is essential for AI for making better decisions and plays a critical role in forecasting. As the data is collected from different sources, there is also a hindrance in accessing the data [27]. So, organizations have to put their work into this area to overcome this difficulty.

6.3.2.7 Interpretability and trust

For an organization, it is easier to convince their stakeholders that specific processes or solutions have to be adopted to improve the processes backed up by expert domains. Explainable AI emerges as one of the leading researches to direct AI adoption in industry interoperability [29]. Interpretability is highly required in decision making where the decision is made from the insight offered by the AI algorithm. This also is very helpful in predicting maintenance and predicting quality which are high-risk applications. In these high-stake applications, explainability and interpretability are anticipated to minimize the risk and increase users' faith [31].

6.3.2.8 Data auditing

The data is composed from various sources, which results in a massive amount of unstructured and assorted data. So, the labeling of data is essential, which helps identify whether the data we have is suitable for the application. This auditing should be done in real-time with a low latency rate, a barrier for industrial AI [31].

6.4 BIG DATA

Big data is a way to analyze and extract information from a large set of data that becomes complex to deal with traditional methods. Big data is generally associated with 3V: variety, velocity, and volume. Big data can achieve the variety, velocity, volume, and criticality of data to process the workload in real-time [31]. Big data is sufficient to capture and handle the data in time and with tremendous value that may not be handled with traditional practices. Big data is highly used in predictive analysis, user behavioral analytics, or other analytics which use massive data. It also helps find the trends in businesses, diseases, criminal cases, etc. Big data integrates the technologies that help enlighten the insight of complex, diverse, and massive-scale data. To manage this data, we can use big data technology. Coming data may be structured or unstructured, but big data manages these data in real-time and gives the advantage of making the right decisions with high precision.

6.4.1 Big data in industry

Big data can be used in the healthcare departments. Knowing a variety of available data sources is crucial to finding out the actual effectiveness of big data. While designing the architecture for the data generated within the system have both functional and non-functional constraints which are generally associated with service accessibility, safety, real-time constraints, and integration of different devices. There is a lack of reference architecture that guided the adaptation of the level of information which led for the researchers to develop the architecture for Industry 4.0 that can assist the developers in developing an extensible, easily scalable, distributed, and flexible environment for Industry 4.0 [32]. Data can be collected from online social networking, which helps the practitioners review the patients' records and share medical information [33]. Big data is also used to gather data for monitoring and maintenance in the manufacturing industry [34]. Big data is also used for extracting the information from the semiconductor management that is one of the complex processes. Khakifirooz et al. [35] proposed a framework that combined a Bayesian approach for complex semiconductor manufacturing to extract information from its data. This framework can also handle several factors of collinearity and high dimensionality. Sahal et al. [36] map the big data for predictive maintenance use requirements. This will be helpful for decision-making, data analytics, and developers.

For self-learning and data management of the machine, big data is used. Cyber-physical systems (CPS) are the main constituent of Industry 4.0, which helps automate the existing industrial operations and processes. Vertical integration of different components within the industry is essential for adopting flexibility and reconfiguration of manufacturing systems which eventually creates the intelligent factory. Decision-making ability and distributed cooperation between different nodes lead to a high level of flexibility. To achieve this, feedback is required, which assists the main coordinators in coordinating and helps in achieving high effectiveness. In those cases, the feedback system with big data assistance can improve efficiency [37]. Big data with a high level of abstraction makes it possible to query the monitored data. It also makes it possible to incorporate data on-the-fly semantics in visualizing results [38]. With the assistance of 5V features of big data, it is possible to construct the data in systematic form and provide comprehensive information by data mining and knowledge discovery. Big data is also applicable for predicting maintenance with the help of the information collected about equipment design failure, staff behavior, and workers' habits, product design defects, etc. [39].

Big data also finds its application in the smart grid as it provides the data of changing demand of electricity quickly and efficiently. It also provides observing and concurrent control of unforeseen changes and ensures the quality of service [40]. Data management and distribution in big data are vital for achieving self-awareness and self-learning. In addition, flexibility and capabilities supported by cloud computing are predetermined. We still have to overcome the challenges of adopting a healthy algorithm that we can implement effectively on the current management technologies. We can use the appropriate sensor installations to extract information such as pressure, temperature, etc. We can also harvest the historical data by using data mining techniques. Transforming nodes consist of several components like predictive analytics, a visualization tool, and an integrated platform. These platforms can be chosen based upon the application. These tools, when integrated, give fruitful results to the users that they can utilize for making a better decision and predicting the forthcoming [41]. Advancement in technologies enables organizations to optimize the processes and plant parameters effectively. The primary purpose of organizations is to improve productivity, increase output or reduce cost. With the help of concurrent data capability and processing capabilities with progressive algorithms enables a profit per hour approach. This approach helps organizations to optimize the available resources. Big data algorithms are a critical enabler to this approach [42].

Smart grid for Industry 4.0 provides the dependable and efficient conveying of the information gathered from the IoT (Internet of Things), empowering the CPS (cyber-physical system) such as sensors placed in an isolated location. Smart grids have an extremely harsh environment that causes the multipath effect, heat, signal fading, high noise, and electromagnetic intrusion, which eventually affect the signal and trigger error in the multichannel wireless sensor networks. So, it becomes essential to observe and control the unexpected power generation changes and distribution in real-time with high efficiency, enabling good quality of service. Big data provides an efficient way to monitor and control unexpected changes with the help of data collected from different sources via the Internet of Things [43].

6.4.2 Challenges for big data implementation

To implement big data, the organization has to face numerous challenges. A few of these challenges are discussed below:

6.4.2.1 Complexity in data

Data management complexity will create complexity in the data and technology like big data. Larger the complexity, the more significant the uncertainty of extracting the valuable information from the complex data set. For long term management of big data a large amount of assessment is required for the long term. The strategies for fragments and frameworks that work with big data will need understanding of both innovations and the necessity of consumers that handle the issues being inquired which means not all big data and its precondition are the same [44].

6.4.2.2 Data quality

For training the model, the dataset required should be clean with a minimum bias factor. If the model is trained on biased and unfiltered data, then the results will have more flaws [18] – the more superb the quality of the data, the more accurate the results. Not having access to high-quality data sets will lead to substandard AI systems. Also, transparency in the product quality can be supervised with the help of AI-based machine learning, which helps detect outliers in the dataset and provide a standard quality [27]. Data quality can be characterized into four main dimensions: intrinsic, contextual, representational, and accessibility. Minimal research has been carried out for assessing these dimensions, which eventually act as a barrier [30].

6.4.2.3 Data accessibility

Data accessibility is essential for extensive data analysis and plays a critical role in forecasting. As the data is collected from different sources, there is also a hindrance in accessing the data. So, organizations have to put their work into this area to overcome this difficulty. Few pieces of research have already been done in this area, which tries to gain trust while connecting with unauthorized nodes in the industrial internet of things. We also have to consider the complexity in cryptographic and its influence on the organizational data and convenience needed to be considered and make sure that the legitimacy of Industrial Internet of Things (IIoT) nodes in the organizational environment of smart devices [30].

6.4.2.4 Data auditing

The data is composed from various sources, which results in a massive amount of unstructured and various data. So, data labeling is essential, which helps identify whether the data we have is suitable for the application. This auditing should be done in real-time with a low latency rate, a barrier for industrial AI [27]. Fan et al. [45] proposed a decentralized auditing method that collects the data from the industrial surrounding. He disintegrated the auditing into three different stages: public, dynamic, and batch auditing, which brings various advantages over the traditional auditing approach. Wu et al. [46] proposed a method to integrate the blockchain and edge computing into a scalable and safe data management scheme.

6.4.2.5 Data transmission

The data is collected from different sources, and those sources have different communication channels which work on different standards. Due to the lack of unanimous standards for communication and data transfer, organizations face difficulty transferring the data. In this scenario, interoperability should be there among different communication channels. Cotel et al. [47] proposed methods which provide quick, secure, and efficient means of transmitting data from advanced sensors to cloud servers. He collected the data from waste organization sources form industries, which helps in decluttering the data from the intelligent factories in efficient ways. Remote maintenance is becoming a new trend in the industry, which requires data transfer in a much more efficient way and utilizes technologies like augmented reality for self-maintenance of factories and minimizes the task for operators' managements [48].

6.4.2.6 Poisoning attacks

These kinds of attacks are very difficult to find out. In these attacks, the data label is manipulated, which then passes through the algorithm. The algorithm shows results based on wrongly labeled data and reduces the performance of the implemented model [27].

6.4.2.7 Trojan attacks

In this type of attack, the attackers try to manipulate the weights of the machine learning algorithms while maintaining the similar structure. This kind of attack is difficult to identify as it operates in normal conditions and triggers only from specific inputs and can cause huge damage to organizations [49].

6.4.2.8 Model extraction and stealing attacks

Model extraction and stealing attacks are associated with the reconstruction or cloning of the target model to negotiate the properties and characteristics of the training dataset [50]. In other attacks, knowledge of the machine learning algorithm's architecture is required, but knowledge of those ML algorithm's architecture or trained datasets is not required in this type of attack. The invader invasion in the structure is assumed to have entrance beside with queries submitted replies [51, 52].

6.5 CONCLUSION

Industry 4.0 is heavily influenced by advanced technologies. Blockchain provides a platform over which information can be shared with a high level of trust. Information shared will be stored with all the participating members, improving the collaboration among the partners. Blockchain also tracks the transactions and is immutable, which is one of the main reasons this technology is used for financial transactions. It also provides decentralization, encryption for the ledger of computer filling, to meet the requirement of Industry 4.0, i.e., automation. Artificial intelligence is one of the critical technologies, which helps in decision-making with a high level of accuracy. Artificial intelligence is an essential technology for Industry 4.0 to be fully functional. Industry 4.0 is strongly dependent on automation and less intervention by humans. To achieve that, we need to have good artificial intelligence for training the machine with machine learning. So, it becomes essential for companies to spend their money on bringing and implementing this technology in their industry. For making a better decision, an enormous amount of data is required to be communicated in real-time. For this purpose, big data technology is a critical technology that helps in providing the data in real-time. Big data provides a massive amount of high-velocity data and provides variable data. This chapter presented the impact of blockchain, artificial intelligence, and big data on Industry 4.0 and their challenges.

REFERENCES

1. Wright, A. and De Filippi, P., 2015. Decentralized blockchain technology and the rise of lex cryptographia. Available at SSRN 2580664.
2. Catalini, C. and Gans, J.S., 2020. Some simple economics of the blockchain. *Communications of the ACM*, 63(7), pp. 80–90.
3. Gupta, R., Tanwar, S., Kumar, N. and Tyagi, S., 2020. Blockchain-based security attack resilience schemes for autonomous vehicles in industry 4.0: A systematic review. *Computers & Electrical Engineering*, 86, p. 106717.
4. Mehta, D., Tanwar, S., Bodkhe, U., Shukla, A. and Kumar, N., 2021. Blockchain-based royalty contract transactions scheme for industry 4.0 supply-chain management. *Information Processing & Management*, 58(4), p. 102586.
5. Liu, X.L., Wang, W.M., Guo, H., Barenji, A.V., Li, Z. and Huang, G.Q., 2020. Industrial blockchain based framework for product lifecycle management in industry 4.0. *Robotics and Computer-Integrated Manufacturing*, 63, p. 101897.

6. Lin, C., He, D., Huang, X., Choo, K.K.R. and Vasilakos, A.V., 2018. BSeIn: A blockchain-based secure mutual authentication with fine-grained access control system for industry 4.0. *Journal of Network and Computer Applications*, 116, pp. 42–52.

7. Rathee, G., Balasaraswathi, M., Chandran, K.P., Gupta, S.D. and Boopathi, C.S., 2021. A secure IoT sensors communication in industry 4.0 using blockchain technology. *Journal of Ambient Intelligence and Humanized Computing*, 12(1), pp. 533–545.

8. Javaid, M., Haleem, A., Singh, R.P., Khan, S. and Suman, R., 2021. Blockchain technology applications for industry 4.0: A literature-based review. *Blockchain: Research and Applications*, 2, p. 100027.

9. Rane, S.B. and Narvel, Y.A.M., 2019. Re-designing the business organization using disruptive innovations based on blockchain-IoT integrated architecture for improving agility in future Industry 4.0. *Benchmarking: An International Journal Vol. 28 No. 5, pp.* 1883–1908.

10. Pinheiro, P., Macedo, M., Barbosa, R., Santos, R. and Novais, P., 2018, June. Multi-agent systems approach to industry 4.0: Enabling collaboration considering a blockchain for knowledge representation. In *International Conference on Practical Applications of Agents and Multi-Agent Systems* (pp. 149–160). Springer, Cham.

11. Viriyasitavat, W., Da Xu, L., Bi, Z. and Sapsomboon, A., 2020. Blockchain-based business process management (BPM) framework for service composition in industry 4.0. *Journal of Intelligent Manufacturing*, 31(7), pp. 1737–1748.

12. Khan, P.W., Byun, Y.C. and Park, N., 2020. IoT-blockchain enabled optimized provenance system for food industry 4.0 using advanced deep learning. *Sensors*, 20(10), p. 2990.

13. Lee, J., Azamfar, M. and Singh, J., 2019. A blockchain enabled Cyber-Physical System architecture for Industry 4.0 manufacturing systems. *Manufacturing letters*, 20, pp. 34–39.

14. Russell, S. and Norvig, P., 2002. *Artificial Intelligence: A Modern Approach*, Pearson Education, Inc., Upper Saddle River, New Jersey 07458.

15. Nilsson, N.J. and Nilsson, N.J., 1998. *Artificial Intelligence: A New Synthesis*. Morgan Kaufmann Publishers, Inc. San Francisco, CA 94104-3205.

16. Luger, G.F., 2005. *Artificial Intelligence: Structures and Strategies for Complex Problem Solving*. Pearson Education, Inc, Boston, MA 02116.

17. Alvarado, M., 1999. Reseña de" Computational Intelligence: a Logical Approach" In: de David Poole, Alan Mackworth y Randy Goebel. *Computación y Sistemas*, 2(2-3), pp.146–149

18. Lee, J., Davari, H., Singh, J. and Pandhare, V., 2018. Industrial artificial intelligence for industry 4.0-based manufacturing systems. *Manufacturing Letters*, 18, pp. 20–23.

19. Merayo, D., Rodriguez-Prieto, A. and Camacho, A.M., 2019. Comparative analysis of artificial intelligence techniques for material selection applied to manufacturing in industry 4.0. *Procedia Manufacturing*, 41, pp. 42–49.

20. Kebisek, M., Tanuska, P., Spendla, L., Kotianova, J. and Strelec, P., 2020. Artificial intelligence platform proposal for paint structure quality prediction within the industry 4.0 concept. *IFAC-PapersOnLine*, 53(2), pp. 11168–11174.

21. O'Donovan, P., Gallagher, C., Leahy, K. and O'Sullivan, D.T., 2019. A comparison of fog and cloud computing cyber-physical interfaces for industry 4.0 real-time embedded machine learning engineering applications. *Computers in industry*, 110, pp. 12–35.

22. Chen, Y. and Li, Y., 2019. Intelligent autonomous pollination for future farming-a micro air vehicle conceptual framework with artificial intelligence and human-in-the-loop. *IEEE Access*, 7, pp. 119706–119717.

23. Samarasinghe, K.R. and Medis, A., 2020. Artificial intelligence based strategic human resource management (AISHRM) for industry 4.0. *Global Journal of Management and Business Research* 20(2), pp. 7–13.

24. Lee, S.M., Lee, D. and Kim, Y.S., 2019. The quality management ecosystem for predictive maintenance in the industry 4.0 era. *International Journal of Quality Innovation*, 5(1), pp. 1–11.

25. Kipper, L.M., Furstenau, L.B., Hoppe, D., Frozza, R. and Iepsen, S., 2020. Scopus scientific mapping production in industry 4.0 (2011–2018): A bibliometric analysis. *International Journal of Production Research*, 58(6), pp. 1605–1627.

26. Strozzi, F., Colicchia, C., Creazza, A. and Noè, C., 2017. Literature review on the 'smart factory' concept using bibliometric tools. *International Journal of Production Research*, 55(22), pp. 6572–6591.

27. Jagatheesaperumal, S.K., Rahouti, M., Ahmad, K., Al-Fuqaha, A. and Guizani, M., 2021. The duo of artificial intelligence and big data for industry 4.0: Review of applications, techniques, challenges, and future research directions. arXiv preprint arXiv:2104.02425.

28. Lee, J., Azamfar, M. and Singh, J., 2019. A blockchain enabled cyber-physical system architecture for industry 4.0 manufacturing systems. *Manufacturing Letters*, 20, pp. 34–39.

29. Parente, M., Figueira, G., Amorim, P. and Marques, A., 2020. Production scheduling in the context of industry 4.0: Review and trends. *International Journal of Production Research*, 58(17), pp. 5401–5431.

30. Peres, R.S., Jia, X., Lee, J., Sun, K., Colombo, A.W. and Barata, J., 2020. Industrial artificial intelligence in industry 4.0-systematic review, challenges and outlook. *IEEE Access*, 8, pp. 220121–220139.

31. Velásquez, N., Estevez, E. and Pesado, P., 2018. Cloud computing, big data and the industry 4.0 reference architectures. *Journal of Computer Science and Technology*, 18(3), pp. e29–e29.

32. Martínez, P.L., Dintén, R., Drake, J.M. and Zorrilla, M., 2021. A big data-centric architecture metamodel for industry 4.0. *Future Generation Computer Systems*, 125, pp. 263–284.

33. Aceto, G., Persico, V. and Pescapé, A., 2020. Industry 4.0 and health: Internet of things, big data, and cloud computing for healthcare 4.0. *Journal of Industrial Information Integration*, 18, p. 100129.

34. Faheem, M., Butt, R.A., Ali, R., Raza, B., Ngadi, M.A. and Gungor, V.C., 2021. CBI4. 0: A cross-layer approach for big data gathering for active monitoring and maintenance in the manufacturing industry 4.0. *Journal of Industrial Information Integration*, 24, p. 100236.

35. Khakifirooz, M., Chien, C.F. and Chen, Y.J., 2018. Bayesian inference for mining semiconductor manufacturing big data for yield enhancement and smart production to empower industry 4.0. *Applied Soft Computing*, 68, pp. 990–999.

36. Sahal, R., Breslin, J.G. and Ali, M.I., 2020. Big data and stream processing platforms for industry 4.0 requirements mapping for a predictive maintenance use case. *Journal of Manufacturing Systems*, 54, pp. 138–151.

37. Wang, S., Wan, J., Zhang, D., Li, D. and Zhang, C., 2016. Towards smart factory for industry 4.0: A self-organized multi-agent system with big data-based feedback and coordination. *Computer Networks*, 101, pp. 158–168.

38. Berges, I., Ramírez-Durán, V.J. and Illarramendi, A., 2021. A semantic approach for big data exploration in industry 4.0. *Big Data Research*, 25, p. 100222.

39. Yan, J., Meng, Y., Lu, L. and Li, L., 2017. Industrial big data in an industry 4.0 environment: Challenges, schemes, and applications for predictive maintenance. *IEEE Access*, 5, pp. 23484–23491.

40. Faheem, M., Fizza, G., Ashraf, M.W., Butt, R.A., Ngadi, M.A. and Gungor, V.C., 2021. Big data acquired by internet of things-enabled industrial multi-channel wireless sensors networks for active monitoring and control in the smart grid industry 4.0. *Data in Brief*, 35, p. 106854.

41. Lee, J., Kao, H.A. and Yang, S., 2014. Service innovation and smart analytics for industry 4.0 and big data environment. *Procedia CIRP*, 16, pp. 3–8.

42. Hammer, M., Somers, K., Karre, H. and Ramsauer, C., 2017. Profit per hour as a target process control parameter for manufacturing systems enabled by big data analytics and industry 4.0 infrastructure. *Procedia CIRP*, 63, pp.715–720.

43. Faheem, M., Fizza, G., Ashraf, M.W., Butt, R.A., Ngadi, M.A. and Gungor, V.C., 2021. Big data acquired by internet of things-enabled industrial multi-channel wireless sensors networks for active monitoring and control in the smart grid industry 4.0. *Data in Brief*, 35, p. 106854.

44. Muhammad, N.K., Osman, N.H. and Salleh, N.A., 2021, July. A conceptual framework for big data analytics adoption towards the success of industry 4.0. In *Business Innovation and Engineering Conference 2020* (BIEC 2020) (pp. 184–187). Atlantis Press, Malaysia.

45. Fan, K., Bao, Z., Liu, M., Vasilakos, A.V. and Shi, W., 2020. Dredas: Decentralized, reliable and efficient remote outsourced data auditing scheme with blockchain smart contract for industrial IoT. *Future Generation Computer Systems*, 110, pp. 665–674.

46. Wu, Y., Dai, H.N. and Wang, H., 2020. Convergence of blockchain and edge computing for secure and scalable IIoT critical infrastructures in industry 4.0. *IEEE Internet of Things Journal*, 8(4), pp. 2300–2317.

47. Cotet, C.E., Deac, G.C., Deac, C.N. and Popa, C.L., 2020. An innovative industry 4.0 cloud data transfer method for an automated waste collection system. *Sustainability*, 12(5), p. 1839.

48. Silvestri, L., Forcina, A., Introna, V., Santolamazza, A. and Cesarotti, V., 2020. Maintenance transformation through Industry 4.0 technologies: A systematic literature review. *Computers in Industry*, 123, p. 103335.

49. Liu, Y., Mondal, A., Chakraborty, A., Zuzak, M., Jacobsen, N., Xing, D. and Srivastava, A., 2020, March. A survey on neural trojans. In *2020 21st International Symposium on Quality Electronic Design (ISQED)* (pp. 33–39). IEEE, Santa Clara, CA, USA, 2020.

50. Juuti, M., Szyller, S., Marchal, S. and Asokan, N., 2019, June. PRADA: Protecting against DNN model stealing attacks. In *2019 IEEE European Symposium on Security and Privacy (EuroS&P)* (pp. 512–527). IEEE, Stockholm, Sweden.

51. Yang, C., Kortylewski, A., Xie, C., Cao, Y. and Yuille, A., 2020, August. Patchattack: A black-box texture-based attack with reinforcement learning. In *European Conference on Computer Vision* (pp. 681–698). Springer, Cham.

52. Xu, H., Ma, Y., Liu, H.C., Deb, D., Liu, H., Tang, J.L. and Jain, A.K., 2020. Adversarial attacks and defenses in images, graphs and text: A review. *International Journal of Automation and Computing*, 17(2), pp. 151–178.

Chapter 7

Enhanced sensor and improved connectivity as key enablers of Industry 4.0

Rupen Trehan, Kuldip Singh Sangwan, Perminderjit Singh, and Narpat Ram Sangwa

7.1 INTRODUCTION

Industry 4.0 has emerged primarily due to the emergence of digital technologies connecting the physical and digital world thereby providing better connections and meaningful collaborations among partners, suppliers, people, and processes. Industry 4.0 enhances automation establishing full-fledged machine-to-machine communication for self-monitoring, self-communication, self-diagnose, self-control, and self-production [1]. Smart sensors and enhanced communication make it possible for a machine or system to diagnose and analyze the issues without human interventions [2]. The four basic characteristics or requirements of an Industry 4.0 system – connected machines, self-learning and self-optimization, real-time monitoring and control, and autonomous production are fully filled with the new generation of cost-effective sensors and communication capabilities [3]. The term Industry 4.0 was coined by representatives from different fields in 2011 in Germany. It is an umbrella term referring to the changes happening in the manufacturing value chain, powered by emerging technologies encompassing the different aspects of industrial production. Industry 4.0 is also known by different names like smart manufacturing and digital manufacturing in the USA and India, respectively [3]. Without emphasising the naming, the root focus remains on connectivity, automation, machine learning, and real-time data [4].

For the system to be intellectual – self-learning and self-adaptation – sensors are important. The sensors are categorised based on their suitability and use. Some sensors are mass-produced and available in the market at reasonable prices. Pressure sensors, position sensors, force sensors, temperature sensors, and flow sensors are the most common sensors. There is hardly any system or product which does not have sensors; from agriculture to aerospace, daily life to defence, bicycle to car, hospital to hospitality, etc. Multiple transitions can be found by using sensors. Also, some of the internal dislocations go through the stages of identification, measurement, analysis, and processing. Improvement in sustainability can be done with the tracking of real-time output by smart factories and also potential plant

DOI: 10.1201/9781003246466-7

maintenance expenses can be reduced with the help of automated control systems [5]. Digitalization may boost production mobility, giving advanced manufacturing companies a competitive advantage. Installation of the sensor may grow in digital supply chains, automated production lines, and process management. The Industry 4.0 paradigm plays an important role in the designing of the robotic platform [6].

With the introduction of enabling technologies like additive manufacturing, the Internet of Things, cloud computing, and big data, Industry 4.0 has drastically transformed the processes and systems of production. Sensors are critical in this regard, for extracting data on environmental conditions, production, equipment health, and spare parts, all of which are vital for enhancing many areas of industrial processes [7]. Installation of sensors throughout the factory to monitor physical measurements such as deformation, temperature, and vibrations might impact production in real-time. Various researchers have identified many critical success factors for the successful adoption of Industry 4.0 [8]. This chapter presents two important critical success factors of Industry 4.0, namely enhanced sensors and improved connectivity.

7.2 ENHANCED SENSORS

The following characteristics of sensors drive the successful adoption of Industry 4.0 by organizations. A classification of the enhanced sensor is shown in Figure 7.1.

7.2.1 Versatility

Heterogeneity of specific sensors is required for sensing temperature, acceleration, gas concentration, or tire slip in the automotive industry. A system

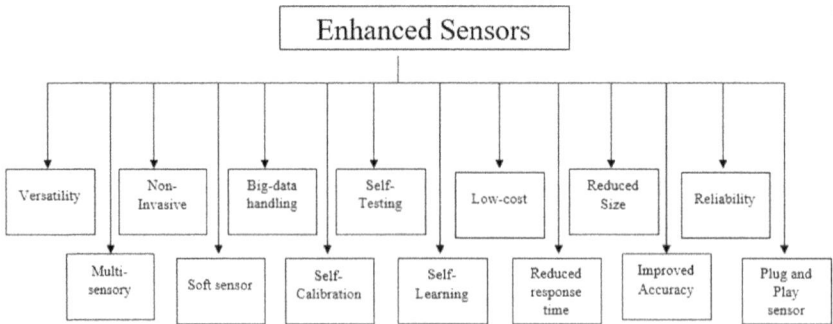

Figure 7.1 Classification of enhanced sensors

is required to synchronize, coordinate, and store measuring data to assess sensor signals efficiently. Due to the versatility of software and hardware design, the application of the systems is not limited to a single industry [9]. These versatile sensors can measure the biomass concentration density of microalgae by sending optical density and biomass concentration to other devices [10]. The total integration or customizable micro fluidic control or sensor, which can be used in laboratory and chemical control and sensing areas, which can be manipulated by an on-board microcontroller [11]. Similarly, versatile sensors are available, which can measure temperature, capacitance resistance with minimum power usage and small chip area [12]. This shows that the versatile sensor is capable of sensing data across all industries. Designing these sensors is a crucial step that can affect the whole process.

7.2.2 Multisensory

Multisensory systems are gaining more popularity due to their advantages like enhancing the system robustness and reliability and improving data credibility and data information efficiency. Now, the application of these sensors is not limited to civil industries, but their applications are seen in military fields also [13]. With the help of a multisensory, the data is collected in a multi-scale for predicting the outcomes. The multi-scale methods are more effective than the non-multi-scale methods. Jiang et al. [14] suggested that having the multi-scale in different layers will improve the system's effectiveness. Deep multisensory learning is a more advanced approach than other learning approaches that improve the effectiveness and stability of the multisensory [15]. Multi-sensors also provide real-time monitoring of the systems, which helps to make better decisions [16]. Li et al. [17] presented PCB detection multisensory for the polarization and infrared imaging to detect the defect even in low lighting conditions.

7.2.3 Non-invasive

Non-invasive sensors are gaining popularity because they reduce pain, are suitable to wear, and able to monitor continuously in real-time. Today, many wearable sensors are lightweight, stretchable, flexible, and biocompatible with the skin [18]. For a glaucoma treatment, intraocular pressure monitoring is essential, and a wearable non-invasive contact lens helps detect intraocular pressure [19]. Non-invasive sensors have many applications in the medical industry to measure and monitor the different parameters for diagnosing different diseases. These sensors give the advantage of blood-free diagnosis and prevent the earlier risk associated with the traditional invasive method. A non-intrusive monitoring approach reduces the complexity and cost involved with multi-sensor arrangements [19].

7.2.4 Soft/virtual sensor

Sangwan et al. [20] presented a non-intrusive smart energy soft sensor to quantify a CNC machining centre's energy consumption at the unit level and suggested the potential areas of energy and time-saving. The soft sensors mitigate the complexity associated with reconfiguration and installation of hard sensors. A soft or virtual sensor is software capable of measuring and processing. For the betterment of the processes, complete monitoring is required. Some quantities are difficult to measure. To overcome this; a data-driven soft sensor comes into the picture. The real-time finding of free lime has promises for quality control [21]. Huang et al. [22] proposed a deep learning technique for a data-driven soft sensor for measurement, processing, and computation. The results show better performances as compared to the traditional methods. Monitoring energy consumption in an ineffective way leads towards energy conservation.

7.2.5 Big-data handling

As most of the machinery is connected to the sensors, the amounts of data that flow through the sensor network have increased significantly. Yamazaki et al. [23] proposed a sensor that can handle the massive amount of disordered data and map that data to offset the sparse bitmap method. The results were compared with conventional indexing methods, which show that the new sensor offers 12 times the writing speed than the existing dense indexing methods. IoT applications are used for analysing the dynamic environment properties by using geo-sensors. The measured data sensed by these sensors are not interoperable with other devices, and a huge amount of data is shared in real-time. Kim and Jang [24] proposed efficient in-memory processing for a huge amount of data for heterogeneous geo-sensors, which helps in reducing the insertion and searching time and improves the memory usage for huge data. The widespread development of machine learning algorithms is providing an added advantage for the handling and analysis of big data. The recent research in data mining, data cleaning, data exploration, and data visualization in conjunction with feature extraction techniques are helping organizations in better decision-making.

7.2.6 Self-calibration

Self-calibration is performed for continuously finding the position of the receiver and transmitter with minimum offset. Ferranti et al. [25] used the algebraic approach as an iterative approach showing that the slight noise and initialization effect converge. An algebraic approach is a powerful tool that helps overcome the iterative approach's challenges. Elangovan et al. [26] used a circuit capable of self-calibration with advantages like independence from the capacitor and more negligible effect due to the non-ideal ties in the circuit leading to lower error. Hong et al. [27] proposed a method where the array calibration is classified into three steps. In the first step, the intrinsic variable of the sensor was found independently. In the second step, the coordinate of the sensor was computed with respect to the reference. In the final step, the two stages of integration take place. The results show considerable improvement in accuracy and precision.

7.2.7 Self-testing

Shen et al. [28] designed and fabricated a sensor, which is a 2D intelligent flow sensor with self-testing functionality having 1.414 times more magnitude than the existing sensors. Dong et al. [29] proposed a method for self-temperature-testing by analyzing different temperature factors when the quartz resonator vibrates in its mode. Beat frequency temperature can be sensed without the use of actual temperature sensors, which enables the sensor to measure the temperature automatically. Testing and calibration consume a considerable time and cost in real-time monitoring of devices, and this is further aggravated if the device is difficult to reach. Self-testing and self-calibration not only improve the data accuracy and decrease the associated costs, but are also a necessity in real-time data acquisition for Industry 4.0 scenarios.

7.2.8 Self-learning

Self-learning is the technique of acquiring net information and updating it over time. A self-learning sensor enables the child sensor to coordinate its wake up with parent sensors without sharing information. This helps in saving energy and reducing the latency problem. Dinh et al. [30] proposed a self-learning sensor having considerable improvement in energy efficiency and packet latency reduction. Most of the systems are now connected to the Internet, and the radio band is populated with lots of unlicensed bands. Chincoli and Liotta [31] demonstrated the application of machine learning to reduce the level of power consumption associated with wireless sensor networks by developing a protocol through the reinforcement learning process with a multi-agent system. This system reduces energy consumption and packet delay more significantly than traditional methods.

7.2.9 Low-cost

Linh et al. [32] provided a low-cost sensor for urban air pollution sensing. The primary driver of the low-cost sensor is to provide high-density spatio-temporal pollution data and identify the hotspots with any further financial investment. Energy consumption is a significant cost in the case of wireless sensor networks. Huang et al. [33] reduced the communication energy cost by 88% of the cluster by integrating the predictor with Kalman filter

(KF). The soft sensors provide cost-effective solutions. The low-cost sensors have created a possibility of Industry 4.0 adoptions in cash-strapped micro, small, and medium enterprises. The low-cost sensors provide an opportunity for the adoption of Industry 4.0 in poor and emerging countries like India.

7.2.10 Reduced response time

In today's world, highly responsive and sensitive sensors are required for the industry. A sensor prepared by hydrothermal method with relevant sensing mechanism of TiO2 film-based hydrogen sensor that shows brilliant performance. The sensor detects the concentration limit as low as 1 ppm in the response time of nine seconds [34]. Wen et al. [35] investigated the response time of the microfiber sensor to measure the temperature in a liquid atmosphere and compared the results with fibre Bragg grating. The result showed that the microfiber sensor response time is 13.1ms, two times lesser than the response time for fibre Bragg grating.

7.2.11 Reduced size (miniaturization)

As the technologies evolve, we require new sensors with small sizes, which can perform the work effectively. We have seen how the size of memory storage devices has reduced over a decade or so. The study developed the flexible pressure sensor based on the nano hierarchy structure. The sensitivity of the sensor is high (~ 0.055 kPa^{-1}). Response time and relaxation time are also reduced due to the hierarchical structure in the capacitive sensor [36]. New age nano and micro fabrication techniques can miniaturize the sensors without compromising sensing performance [37]. The miniaturization of sensors has improved the adoption rate as it is convenient to add to the machines and tools.

7.2.12 Improved accuracy

Global Positioning Systems (GPS) have problems in indoor applications. An alternate solution is Zig Loc, a system that does not require additional infrastructure based on Wi-Fi hotspots. The sensor node routines the ZigBee (IEEE 802.15.4) section to measure the strength of the received signal (RSS) of the Wi-Fi contact plug signal. It then uses the fingerprints collected for the Wi-Fi localization system to estimate the location of the sensor node's difference between access points and must be the same when they are being measured. Analysis based upon experiments was conducted and the result of the analysis is that by using the differential fingerprinting method; Zig loc measurements can be improvised by 26 percent [38]. Three-dimensional data can be used for analysis and the preparation of documents. Errors or deviations from various sources can pair the data obtained from terrestrial laser scanners (TLSs). A calibration routine schedule is essential for TLSs to maintain high quality. A dense 3D target field was used to self-calibrate the Faro Photon 120 scanner. Approximately four parameters for calibration were concluded from sources. The significant test was performed using the T-test which shows the constant error (8.9 mm) and collimation axis error (-4.3) [39]. Azimirad et al. [40] used dempster-shafer theory to enhance the output by 97 percent.

7.2.13 Reliability

Transmission of data packets must be via different paths because when the same data packets are moved along the same path, the network becomes busier. However, a problem arises in the concept of different path transmission. A concept of the Residue number system is in use for detection and identification of errors and the residue system is based upon a special module facility that is applicable for increment in output [41]. Low-power wireless sensor networks are widely used in IoT-enabled smart applications and infrastructure. However, it results in dependability and security risks due to their low energy consumption and lack of processing capabilities. As a result, sensing devices with lesser capability must be used intelligently and with rigorous technological considerations in order to optimise their possibilities. Sen and Jayawardena [42] provided a new approach for improving reliability and cybersecurity by balancing sensor energy consumption while preserving communication quality. Banerjee et al. [43] showed that wireless sensor networks (WSN) power consumption is directly proportional to its durability, and a more durable WSN ensures greater reliability.

7.2.14 Sensor and actuator integration (plug and play sensor)

Nowadays monitoring systems are expected to be employed with flexible design in terms of maintenance and usability. The modular sensor system is the first air monitoring system with plug and play features with multiple wireless network compatibility, and this system is better than the previously used multiple sensors and modular sensor systems [44]. Interoperability is important for sharing information if they are manufactured by different vendors. For doing so, the application layer protocol should be standardized. These standards should be independent from overlying layers [45]. Chiu et al. [46] proposed a real-time navigation method that can integrate a wide range of sensors while meeting requirements of performance by employing a unique restricted optimum selection process. Hussain and Gurkan [47] found that as more sensors are networked via Internet technologies, network management of sensor nodes is becoming increasingly popular; efforts to standardize sensor networking have resulted in a more interoperable method of sensor identification and transmission. The confluence of sensor networking with internetworking allows for greater flexibility in network management protocol development.

Weddell et al. [48] stated that a new approach allows energy modules to be selected and attached to the sensor node in place without the need for reprogramming or hardware modifications. Sensors can now be powered by ambient energy when it is available, thanks to the recent advancements in energy harvesting technologies. Liu et al. [49] proposed a new platform that includes sensor nodes for monitoring the state of the machines and a gateway unit connecting the nodes to the cloud command and control. The new platform provides enhanced productivity, increased energy efficiency, and lower costs. Martínez et al. [50] suggested that a system for integrating sensors into existing data infrastructures via a plug and play system must be developed so that sensors and observation platforms can be combined as lack of standardization eventually leads to information silos, preventing data from being exchanged efficiently across scientific groups. Gomes et al. [51] mentioned that final IoT users must be able to use these devices without any prior technological understanding.

7.3 IMPROVED CONNECTIVITY

One of the fundamentals of Industry 4.0 is machine-to-machine connectivity to transfer the data across the organization and its suppliers/partners. The various aspects of improved connectivity are as shown in Figure 7.2.

7.3.1 Ad hoc networks

Many mobile nodes get connected for the formation of a mobile Ad hoc network. Infrastructure accidental ways are being used in these imitations.

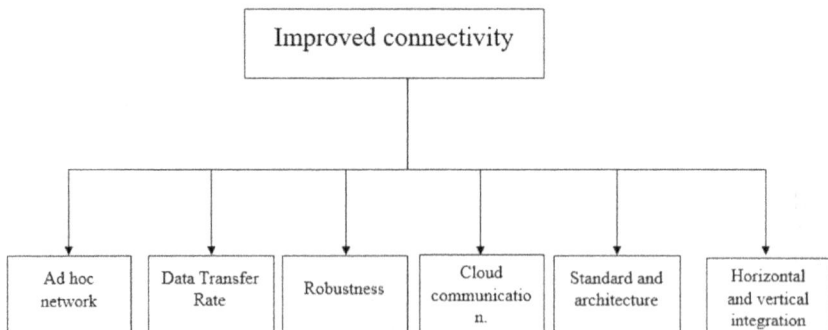

Figure 7.2 Various aspect of improved connectivity

Arbitrary waypoints are modest models that can be useful in some circumstances [52]. Adaptive multipath routing algorithm provides nodes in MAS Nets (Mobile Ad hoc and Sensor Networks) with the highest connectivity life cycle for a particular target. Nodes in MANets can describe the degree of disjoint of the available paths. These nodes execute on-demand route finding and route generation. The suggested multipath routing algorithm is flexible and employs labels to discover routes. Watchdog mechanism is being considered as the process of discovery of route [53]. Hunjet and Hui [54] introduced the concept of distributed MANET (mobile ad hoc network) for relay placement within the nodes. Several algorithms have been developed for MANET. Combined usage of five relays will help to enhance average unit network connectivity by 82 percent. Kamya et al. [55] analyzed the performance of the VANETs (vehicle ad hoc networks) based on connectivity in highly dense environments and found that connectivity probability decreases with interference and increases with communication range. However, analysing the trust level at each node is important as it provides the confidence with which data can be shared from the node [56].

7.3.2 Data transfer rate

The rate at which the data is flowing in a network relay affects the process effectiveness. Yao et al. [57] developed a new approach for incorporating full-duplex messages into powerful wireless power transfer structures. Data and power can be sent via the same inductive connection. This strategy has the potential to reduce the system size as well as cost. However, due to the variety of end systems' capabilities and available network capacity, controlling the pace of bulk data multicast to a wide number of recipients is difficult. Some receivers will lose data if the data transfer rate is too high, necessitating retransmissions, and presented a method for determining the appropriate coil tap location to increase data transmission performance. The system's circuit model is created to analyse data broadcast enactment. Duplexer can be used to split data signals that are sent and received [58]. Ghovanloo and Najafi [59] found that serialization and deserialization channels instead of PCIs (peripheral components interconnect) link double the speed for the same number of PCBs (printed circuit boards). Mittal et al. [60] proposed a method of increasing the adaptability of the protocol for topological changes for VANETs. Rahnama et al. [61] found that serial communications are quickly displacing parallel communications in high-speed data networks and presented a novel connectivity switch solution in High-performance computing) systems. Instead of PCIe (peripheral component interconnect express) standard connections, it employs a novel jitter-free data transport technology for deserialization channels. Sato et al. [62] investigated a delay detection approach to solve issues related to nonlinear STOs (spin-torque oscillators) diffusion and found that the frequencies of

STO as well as the rate of high-data-transfer are important parameters in the frequency transition of STO. Besnoff et al. [63] proposed a communication technique for the wireless power transfer (WPT) at ultra-high data-rate of 3.39 Mb/s, running at 13.56 Megahertz with only an efficiency loss of two percent.

7.3.3 Robustness

Multi-agent topology resilience is important. It's also critical to enhance the robustness without raising the cost. The formation optimization issue is turned into a 0–1 nonlinear programming problem by analysis. The research proposes a genetic method based on chaotic search optimization to solve the issue fast. Deng et al. [64] proposed a 0–1 nonlinear programming model and solved it based on the chaotic search optimization method to improve the network's robustness. Banerjee et al. [65] found that the region-based component decomposition number (RBCDN) helps in identifying the number of connected components in which the network decomposes once all the nodes fail. This is very helpful for finding out the robustness and fault tolerance capability in both wired and wireless networks. Zhu et al. [66] found that allocating connectivity and dependence link together improves robustness increases as compared with the connectivity link or dependence link separately. Thota et al. [67] suggested that an alternate path across the wireless-wired line integrated network may be chosen in the event of a network element failure allowing an untethered wireless device to connect to this network through a dependable high-capacity connection. Abbas et al. [68] proposed an algorithm for improving the structural resilience of networks without adding more links by ensuring that a limited fraction of nodes, known as trustworthy nodes, are always intact and functioning properly. The proposed trusted node robust consensus method showed that resilient consensus can be achieved in sparse networks with few trusted nodes. Ji et al. [69] showed that the robustness of interdependent networks by introducing connection links and link addition techniques outperform current tactics, especially for interdependent networks that can help with resource allocation and topology optimization in an existing interdependent network system.

Zhao et al. [70] examined resilience and connectivity through a broad random graph model of intersection where each node in the graphs is randomly allocated collective items, and two of the nodes have an edge that is undirected if they share one object at least. Jamakovic et al. [71] measured the graph's resilience by node and link connectivity; two topological metrics that indicate the number of nodes and connections that must be deleted in order to disconnect a network. Peng and Wu [72] described that rebuilding the linkages and correcting the node degrees can improve the structural stability of a network, and the ideal network has a topology that is "eggplant-like," along with a cluster of nodes of high-degree at the head and nodes of low-degree dispersed throughout the body.

7.3.4 Cloud communication

It is a data centre hosted service based on demand response management (DRM). It helps to prevent black outs, and is economical to both end-users and energy providers. The advancement of communication networks in modern power grids has enhanced DRM functions and grids are becoming smarter. 5G IoT technologies play an important role in cloud communication. They are focused on the advancement of algorithms like deep learning and machine learning [73].

Sliwa et al. [74] observed that the tremendous increase in machine-type communication develops strain on network set-up. Because of this, more increased the mean data rate by 194 percent and reduced the side-by-side average power consumption by 54 percent by developing resource-efficient channel access approaches to counter system-restricted resource limitation in network structure. Liao et al. [75] provided a new approach for merging the 4G data to upload the data to the cloud server and all the papa metres are coordinated simultaneously. Nair and Chandavarkar [76] studied the Ph value and temperature parameters effects for underwater communication and proposed a routing algorithm based on the SNR (signal to noise ratio). Kimura and Ogura [77] suggested that the Aerial Base Stations (ABS) improve next-generation wireless network coverage by designing a 3D deployment of ABSs. Sliwa et al. [78] mentioned that machine-type communication (MTC) will increase in 5G communication networks, necessitating resource-efficient transmission strategies to maximise system performance. Further they suggested a context-predictive strategy which increased the average rate of data transmission by 194 percent whereas lowered the uplink power consumption by 54 percent. Zhao et al. [79] stated that to satisfy user satisfaction and their effects at the same time is not capable of demand response of single type. For integrated demand response (IDR) develop traditional demand response. An improved communication optimization algorithm was used to solve the communication in quality for IDR.

Wei and Wu [80] suggested a QFD (quality function deployment) approach for the accurate evaluation of the user demand for the communication equipment. Forward cloud resources have become widespread as a result of recent advancements in communication technology, and Car-to-Cloud connections can help with data transfer to the cloud by enabling dynamic learning of fresh map information [81]. Jayalath et al. [82] developed a computing system and cloud storage using the key-value pair architecture and content-based publish/subscribe systems to give an abstract expression.

7.3.5 Standards and architecture

Mengqiu et al. [83] presented a common strategy to standardise the DER (Distributed Energy Resources) system communication network with the use of IEC 61850, which covers communication architecture design and implementation. It may be used to build high-level reasoning agents by plugging in standard reasoning engines which act as plug-in components and talking with the rest of the agents using standard FIPA-based communication protocols. There are three formal reasoning system components that have been constructed and evaluated in combination with a prevailing agent platform called OPAL to implement FIPA (Foundation for Intelligent Physical Agents) standards for communication [83]. Huang et al. [84] suggested two prototype test beds for the DER system at the SMERC laboratory to develop the system architecture configuration, data, and integration of the IEC 61850 standards. Danilin et al. [85] investigated the reconfigurable logic for the address sequence generation, the common task modulation, and filtering for digital communication like mobile terminals and hand-held devices.

7.3.6 Horizontal and vertical integration

Costa et al. [86] investigated the integration of digitalized service/products and commercial models with improved production technologies to empower integration in vertical, horizontal, and endwise forms for efficient digital transformation. This integration is built on modern artificial intelligence solutions for decentralisation of decision-making as well as the safe trustworthy exchange of data throughout the global value chain. Management, forecasting, optimization, and simulation are examples of these efforts, which harmonize the varied properties to satisfy the actual needs [86]. Ye et al. [87] created a dimension data modelling approach based on vertical and horizontal integration. Liu and Wang [88] opined that neural network forecasting accuracy increases with the increase in prediction days for horizontal and vertical integration prediction. Lukoki et al. [89] provided a 12-stepped approach for the implementation of horizontal and vertical integration in cyber physical systems and validated at an assembly line.

7.4 CONCLUSIONS

In this chapter, we study the different sensor characteristics and communication capabilities that can be employed in Industry 4.0 to improve decision-making for better results. The developments in sensor technology, particularly during the last decade, in terms of versatility, non-invasiveness, multisensory, reliability, miniaturization, accuracy, and response time have made the modern-day sensors cost-effective, efficient and effective to be embedded or integrated with machines. The developments in soft/virtual sensors based on machine learning, self-calibration, self-testing, self-learning, and the ability of the sensors to handle big data led the researchers to the cusp of the new paradigm of Industry 4.0 and encouraged the industry to adopt the new paradigm without hesitation. The low cost and plug and play integration of new sensors are expected to increase the penetration of digitalization among SMEs (small and medium enterprises) as these enterprises can hardly afford skilled employees.

Fast, reliable, and robust connectivity by using ad hoc networks and cloud communication with new standards and architecture has made it possible for the industry to employ horizontal and vertical integration across the whole value chain activities. Newer networking methods improve data sharing without loss of data. These methods help in creating a robust network that can work efficiently.

The combination of enhanced sensors and communication capabilities has improved the latency issues thereby making real-time data collection and sharing possible for Industry 4.0 applications. The availability of enhanced sensors and connectivity lay the foundation over which Industry 4.0 related technologies can be implemented without much difficulty.

The list of reviewed papers for enhanced sensor and improved connectivity along with details of journals and publishers are given in Annexure 7.1.

REFERENCES

1. G. Culot, G. Nassimbeni, G. Orzes, and M. Sartor, "Behind the definition of industry 4.0: Analysis and open questions," *Int. J. Prod. Econ.*, 226(August), p. 107617, 2020, doi: 10.1016/j.ijpe.2020.107617.
2. M. Javaid, A. Haleem, R. P. Singh, S. Rab, and R. Suman, "Significance of sensors for industry 4.0: Roles, capabilities, and applications," *Sens. Int.*, 2(June), p. 100110, 2021, doi: 10.1016/j.sintl.2021.100110.
3. M. Sanchez, E. Exposito, and J. Aguilar, "Autonomic computing in manufacturing process coordination in industry 4.0 context," *J. Ind. Inf. Integr.*, 19(July), p. 100159, 2020, doi: 10.1016/j.jii.2020.100159.
4. L. Silvestri, A. Forcina, V. Introna, A. Santolamazza, and V. Cesarotti, "Maintenance transformation through industry 4.0 technologies: A systematic literature review," *Comput. Ind.*, 123, p. 103335, 2020, doi: 10.1016/j.compind.2020.103335.
5. H. C. Müller, A. Hennig, H. Höller, and K. Polster, "Wireless sensors for industry 4.0," *Sensoren und Messsysteme. - 19. ITG/GMA-Fachtagung*, pp. 251–254, 2018.
6. G. M. E-Elahi, "Improvement of traffic overhead and reliability in wireless sensor network," *Int. J. Comput. Sci. Inf. Technol.*, 4(3), pp. 45–59, 2012, doi: 10.5121/ijcsit.2012.4305.
7. J. Sarkar and S. Bhattacharyya, "Application of graphene and graphene-based materials in clean energy-related devices Minghui," *Arch. Thermodyn.*, 33(4), pp. 23–40, 2012, doi: 10.1002/er.

8. T. Wijaya, W. Caesarendra, B. K. Pappachan, T. Tjahjowidodo, A. Wee, and M. I. Roslan, "Robot control and decision making through real-time sensors monitoring and analysis for industry 4.0 implementation on aerospace component manufacturing," *PacRim 2017: IEEE Pacific Rim Conference on Communications, Computers and Signal Processing*, 2017, pp. 1–6, 2017, doi: 10.1109/PACRIM.2017.8121928.

9. M. Lieschnegg, B. Lechner, A. Fuchs, and O. Mariani, "Versatile sensor platform for autonomous sensing in automotive applications," *Int. J. Smart Sens. Intell. Syst.*, 4(3), pp. 496–507, 2011, doi: 10.21307/ijssis-2017-453.

10. R. C. Barbosa, J. Soares, and M. Arêdes Martins, "Low-cost and versatile sensor based on multi-wavelengths for real-time estimation of microalgal biomass concentration in open and closed cultivation systems," *Comput. Electron. Agric.*, 176(July), p. 105641, 2020, doi: 10.1016/j.compag.2020.105641.

11. F. Kehl, V. F. Cretu, and P. A. Willis, "Open-source lab hardware: A versatile microfluidic control and sensor platform," *HardwareX*, 10, p. e00229, 2021, doi: 10.1016/j.ohx.2021.e00229.

12. H. Xin, M. Andraud, P. Baltus, E. Cantatore, and P. Harpe, "A 0.34–571nW all-dynamic versatile sensor interface for temperature, capacitance, and resistance sensing," *ESSCIRC 2019 - IEEE 45th European Solid State Circuits Conference (ESSCIRC)*, pp. 161–164, 2019, doi: 10.1109/ESSCIRC.2019.8902918.

13. B. Shen, Z. Wang, H. Tan, and H. Chen, "Robust fusion filtering over multisensor systems with energy harvesting constraints," *Automatica*, 131, p. 109782, 2021, doi: 10.1016/j.automatica.2021.109782.

14. X. Li, H. Jiang, Y. Liu, T. Wang, and Z. Li, "An integrated deep multiscale feature fusion network for aeroengine remaining useful life prediction with multisensor data," *Knowl. Based Syst.*, 235, p. 107652, 2021, doi: 10.1016/j.knosys.2021.107652.

15. Z. Zheng, A. Ma, L. Zhang, and Y. Zhong, "Deep multisensor learning for missing-modality all-weather mapping," *ISPRS J. Photogramm. Remote Sens.*, 174(March), pp. 254–264, 2021, doi: 10.1016/j.isprsjprs.2020.12.009.

16. L. C. Price, J. Chen, J. Park, and Y. K. Cho, "Multisensor-driven real-time crane monitoring system for blind lift operations: Lessons learned from a case study," *Autom. Constr.*, 124(October), p. 103552, 2021, doi: 10.1016/j.autcon.2021.103552.

17. M. Li, N. Yao, S. Liu, S. Li, Y. Zhao, and S. G. Kong, "Multisensor image fusion for automated detection of defects in printed circuit boards," *IEEE Sens. J.*, 21(20), pp. 23390–23399, 2021, doi: 10.1109/JSEN.2021.3106057.

18. Z. Dou *et al.*, "Wearable contact lens sensor for non-invasive continuous monitoring of intraocular pressure," *Micro Mach.*, 12(2), pp. 1–12, 2021, doi: 10.3390/mi12020108.

19. N. Promphet, J. P. Hinestroza, P. Rattanawaleedirojn, N. Soatthiyanon, K. Siralertmukul, P. Potiyaraj, N. Rodthongkum, "Cotton thread-based wearable sensor for non-invasive simultaneous diagnosis of diabetes and kidney failure," *Sens. Actuators B*, 321(December), p. 128549, 2020, doi: 10.1016/j.snb.2020.128549.

20. N. Sihag, K. S. Sangwan, and S. Pundir, "Development of a structured algorithm to identify the status of a machine tool to improve energy and time efficiencies," *Procedia CIRP*, 69(May), pp. 294–299, 2018, doi: 10.1016/j.procir.2017.11.081.
21. B. Lin, B. Recke, J. K. H. Knudsen, and S. B. Jørgensen, "A systematic approach for soft sensor development," *Comput. Chem. Eng.*, 31(5–6), pp. 419–425, 2007, doi: 10.1016/j.compchemeng.2006.05.030.
22. C. Shang, F. Yang, D. Huang, and W. Lyu, "Data-driven soft sensor development based on deep learning technique," *J. Process Control*, 24(3), pp. 223–233, 2014, doi: 10.1016/j.jprocont.2014.01.012.
23. T. Yamazaki, T. Inoue, H. Sato, N. Takahashi, J. Takagi, and M. Minami, "Efficiently indexing with offset bitmaps for huge sets of slightly disordered sensor data," *2010 8th Asia-Pacific Symposium on Information and Telecommunication Technologies*, APSITT, pp. 2–7, 2010.
24. M. Soo Kim and I. Sung Jang, "Efficient in-memory processing for huge amounts of heterogeneous geo-sensor data," *Spat. Inf. Res.*, 24(3), pp. 313–322, 2016, doi: 10.1007/s41324-016-0029-7.
25. L. Ferranti and K. Astr, "Homotopy continuation for sensor networks," *Proceedings of the European Signal Processing Conference*, pp. 1725–1729, 2021.
26. K. Elangovan and C. S. Anoop, "A self-calibration circuit for the resistance measurement of parallel R-C sensor," *2021 IEEE Second International Conference on Control, Measurement and Instrumentation (CMI 2021)*, pp. 201–205, 2021, doi: 10.1109/CMI50323.2021.9362820.
27. J. H. Hong, D. Kang, and I. J. Kim, "A unified method for robust self-calibration of 3-D field sensor arrays," *IEEE Trans. Instrum. Meas.*, 70, 2021, doi: 10.1109/TIM.2021.3072112.
28. G. P. Shen, M. Qin, and Q. A. Huang, "A cross-type thermal wind sensor with self-testing function," *IEEE Sens. J.*, 10(2), pp. 340–346, 2010, doi: 10.1109/JSEN.2009.2034625.
29. Y. G. Dong, J.-S. Wang, G.-P. Feng, X.-H. Wang, "Self-temperature-testing of the quartz resonant force sensor," *IEEE Trans. Instrum. Meas.*, 48(6), pp. 1038–1040, 1999, doi: 10.1109/19.816110.
30. T. Dinh, Y. Kim, T. Gu, and A. V. Vasilakos, "L-MAC: A wake-up time self-learning MAC protocol for wireless sensor networks," *Comput. Netw.*, 105, pp. 33–46, 2016, doi: 10.1016/j.comnet.2016.05.015.
31. M. Chincoli and A. Liotta, "Self-learning power control in wireless sensor networks," *Sensors (Switzerland)*, 18(2), pp. 1–29, 2018, doi: 10.3390/s18020375.
32. L. T. Phuong Linh, I. Kavalchuk, A. Kolbasov, K. Karpukhin, and A. Terenchenko, "The performance assessment of low-cost air pollution sensor in city and the prospect of the autonomous vehicle for air pollution reduction," *IOP Conf. Ser. Mater. Sci. Eng.*, 819(1), 2020, doi: 10.1088/1757-899X/819/1/012018.
33. Y. Huang, W. Yu, C. Osewold, and A. Garcia-Ortiz, "Analysis of PKF: A communication cost reduction scheme for wireless sensor networks," *IEEE Trans. Wirel. Commun.*, 15(2), pp. 843–856, 2016, doi: 10.1109/TWC.2015.2479234.

34. X. Xia, W. Wu, Z. Wang, Y. Bao, Z. Huang, and Y. Gao, "A hydrogen sensor based on orientation aligned TiO_2 thin films with low concentration detecting limit and short response time," *Sens. Actuators B*, 234, pp. 192–200, 2016, doi: 10.1016/j.snb.2016.04.110.

35. J. H. Wen *et al.*, "Response time of microfiber temperature sensor in liquid environment," *IEEE Sens. J.*, 20(12), pp. 6400–6407, 2020, doi: 10.1109/JSEN.2020.2976535.

36. C. Mahata, H. Algadi, J. Lee, S. Kim, and T. Lee, "Biomimetic-inspired micro-Nano hierarchical structures for capacitive pressure sensor applications," *Meas. J. Int. Meas. Confed.*, 151, p. 107095, 2020, doi: 10.1016/j.measurement.2019.107095.

37. H. Nazemi, A. Joseph, J. Park, and A. Emadi, "Advanced micro-and nano-gas sensor technology: A review," *Sensors (Switzerland)*, 19(6), 2019, doi: 10.3390/s19061285.

38. T. Yamamoto, S. Ishida, K. Izumi, S. Tagashira, and A. Fukuda, "Accuracy improvement in sensor localization system utilizing heterogeneous wireless technologies," *2017 Tenth International Conference on Mobile Computing and Ubiquitous Network (ICMU)*, 2017, pp. 1–6, 2018, doi: 10.23919/ICMU.2017.8330074.

39. M. A. Abbas *et al.*, "Improvement in accuracy for three-dimensional sensor (faro photon 120 scanner)," *Int. J. Comput. Sci. Issues*, 10(1), pp. 176–182, 2013.

40. E. Azimirad and S. R. M. Ghodsinya, "The improvement of uncertainty measurements accuracy in sensor networks based on fuzzy dempster-shafer theory," *Int. J. Adv. Intell. Inform.*, 6(2), pp. 149–160, 2020, doi: 10.26555/ijain.v6i2.461.

41. V. Yatskiv and T. Tsavolyk, "Improvement of data transmission reliability in wireless sensor networks on the basis of residue number system correcting codes using the special module system," *2017 IEEE First Ukraine Conference on Electrical and Computer Engineering (UKRCON)*, pp. 890–893, 2017, doi: 10.1109/UKRCON.2017.8100376.

42. S. Sen and C. Jayawardena, "Reliability and cybersecurity improvement strategies in wireless sensor networks for IoT-enabled smart infrastructures," *2019 Global Conference for Advancement in Technology (GCAT)*, pp. 1–8, 2019, doi: 10.1109/GCAT47503.2019.8978380.

43. A. Banerjee, M. Gavrilas, O. Ivanov, and S. Chattopadhyay, "Reliability improvement and the importance of power consumption optimization in wireless sensor networks," *2015 9th International Symposium on Advanced Topics in Electrical Engineering (ATEE)*, pp. 735–740, 2015, doi: 10.1109/ATEE.2015.7133902.

44. W. Y. Yi, K. S. Leung, and Y. Leung, "A modular plug-and-play sensor system for urban air pollution monitoring: Design, implementation and evaluation," *Sensors (Switzerland)*, 18(1), 2018, doi: 10.3390/s18010007.

45. J. Salazar-Vazquez and A. Mendez-Vazquez, "A plug-and-play hyperspectral Imaging Sensor using low-cost equipment," *HardwareX*, 7, p. e00087, 2020, doi: 10.1016/j.ohx.2019.e00087.

46. H. P. Chiu, X. S. Zhou, L. Carlone, F. Dellaert, S. Samarasekera, and R. Kumar, "Constrained optimal selection for multi-sensor robot navigation using plug-and-play factor graphs," *2014 IEEE International Conference on Robotics and Automation (ICRA)*, pp. 663–670, 2014, doi: 10.1109/ICRA.2014.6906925.
47. S. A. Hussain and D. Gurkan, "Management and plug and play of sensor networks using SNMP," *IEEE Trans. Instrum. Meas.*, 60(5), pp. 1830–1837, 2011, doi: 10.1109/TIM.2011.2113115.
48. A. S. Weddell, N. J. Grabham, N. R. Harris, and N. M. White, "Modular plug-and-play power resources for energy-aware wireless sensor nodes," *2009 6th Annual IEEE Communications Society Conference on Sensor, Mesh and Ad Hoc Communications and Networks*, SECON, 2009, doi: 10.1109/SAHCN.2009.5168947.
49. S. Liu, J. A. Guzzo, L. Zhang, D. W. Smith, J. Lazos, and M. Grossner, "Plug-and-play sensor platform for legacy industrial machine monitoring," in *2016 International Symposium on Flexible Automation (ISFA)*, 2016, pp. 432–435. doi: 10.1109/ISFA.2016.7790202.
50. E. Martínez, D. M. Toma, S. Jirka, and J. Del Río, "Middleware for plug and play integration of heterogeneous sensor resources into the sensor web," *Sensors (Switzerland)*, 17(12), pp. 1–28, 2017, doi: 10.3390/s17122923.
51. J. B. A. Gomes, J. J. P. C. Rodrigues, R. A. L. Rabêlo, S. Tanwar, J. Al-Muhtadi, and S. Kozlov, "A novel Internet of things-based plug-and-play multi gas sensor for environmental monitoring," *Trans. Emerg. Telecommun. Technol.*, 32(6), pp. 1–11, 2021, doi: 10.1002/ett.3967.
52. F. Bai, N. Sadagopan, and A. Helmy, "The important framework for analyzing the impact of mobility on performance of RouTing protocols for AdhocNeTworks," *Ad Hoc Netw.*, 1(4), pp. 383–403, 2003, doi: 10.1016/S1570-8705(03)00040-4.
53. B. Triki, S. Rekhis, and N. Boudriga, "Connectivity-aware and adaptive multipath routing algorithm for mobile adhoc and sensor networks," *Int. J. Wirel. Mob. Netw.*, 7(1), pp. 55–74, 2015, doi: 10.5121/ijwmn.2015.7104.
54. R. Hunjet and P. Hui, "Maintaining connectivity in mobile adhoc networks using distributed optimisation," *2011 Military Communications and Information Systems Conference*, pp. 5–10, 2011, doi: 10.1109/MilCIS.2011.6470392.
55. M. Kamya, J. Mwebaze, and M. Okopa, "Modeling connectivity for vehicular adhoc networks under interference," *2021 Fifth World Conference on Smart Trends in Systems Security and Sustainability (WorldS4)*, pp. 87–95, 2021, doi: 10.1109/WorldS451998.2021.9514056.
56. K. Govindan and P. Mohapatra, "Trust computations and trust dynamics in mobile adhoc networks: A survey," *IEEE Commun. Surv. Tutor.*, 14(2), pp. 279–298, 2012, doi: 10.1109/SURV.2011.042711.00083.
57. Y. Yao, H. Cheng, Y. Wang, J. Mai, K. Lu, and D. Xu, "An FDM-based simultaneous wireless power and data transfer system functioning with high-rate full-duplex communication," *IEEE Trans. Ind. Inform.*, 16(10), pp. 6370–6381, 2020, doi: 10.1109/TII.2020.2967023.

58. S. Bhattacharyya, J. F. Kurose, D. Towsley, and R. Nagarajan, "Efficient rate-controlled bulk data transfer using multiple multicast groups," *IEEE ACM Trans. Netw.*, 11(6), pp. 895–907, 2003, doi: 10.1109/TNET.2003.820247.

59. M. Ghovanloo and K. Najafi, "A high data transfer rate frequency shift keying demodulator chip for the wireless biomedical implants," *Midwest Symp. Circuits Syst.*, 3, pp. 433–436, 2002.

60. S. Mittal, R. Kaur, and K. C. Purohit, "Enhancing the data transfer rate by creating alternative paths for AODV routing protocol in VANET," *2016 2nd International Conference on Advances in Computing, Communication, & Automation (ICACCA) (Fall)*, pp. 7–11, 2016, doi: 10.1109/ICACCAF.2016.7748976.

61. B.Rahnama, A. Sari, and R. Makvandi, "Countering PCIe Gen. 3 data transfer Rate Imperfection using serial data interconnect," *IEEE The International Conference on Technological Advances in Electrical Electronics and Computer Engineering*, 1, pp. 579–582, 2013.

62. R. Sato, K. Kudo, T. Nagasawa, H. Suto, and K. Mizushima, "Simulations and experiments toward high-data-transfer-rate readers composed of a spin-torque oscillator," *IEEE Trans. Magn.*, 48(5 part 1), pp. 1758–1764, 2012, doi: 10.1109/TMAG.2011.2173560.

63. J. Besnoff, M. Abbasi, and D. S. Ricketts, "Ultrahigh-data-rate communication and efficient wireless power transfer at 13.56 MHz," *IEEE Antennas Wirel. Propag. Lett.*, 16, pp. 2634–2637, 2017, doi: 10.1109/LAWP.2017.2736883.

64. Z. H. Deng, J. Xu, Q. Song, B. Hu, T. Wu, and P. Huang, "Robustness of multi-agent formation based on natural connectivity," *Appl. Math. Comput.*, 366, 2020, doi: 10.1016/j.amc.2019.124636.

65. S. Banerjee, S. Shirazipourazad, P. Ghosh, and A. Sen, "Beyond connectivity-new metrics to evaluate robustness of networks," *2011 IEEE 12th International Conference on High Performance Switching and Routing*, HPSR, pp. 171–177, 2011, doi: 10.1109/HPSR.2011.5986022.

66. P. Cui, P. Zhu, K. Wang, P. Xun, and Z. Xia, "Enhancing robustness of inter-dependent network by adding connectivity and dependence links," *Phys. Stat. Mech. Appl.*, 497, pp. 185–197, 2018, doi: 10.1016/j.physa.2017.12.142.

67. S. Thota, P. Bhaumik, P. Chowdhury, B. Mukherjee, and S. Sarkar, "Exploiting wireless connectivity for robustness in WOBAN," *IEEE Netw.*, 27(4), pp. 72–79, 2013, doi: 10.1109/MNET.2013.6574668.

68. W. Abbas, A. Laszka, and X. Koutsoukos, "Improving network connectivity and robustness using trusted nodes with application to resilient consensus," *IEEE Trans. Control Netw. Syst.*, 5(4), pp. 2036–2048, 2018, doi: 10.1109/TCNS.2017.2782486.

69. X. Ji, B. Wang, D. Liu, G. Chen, F. Tang, D. Wei, L. Tu, "Improving interdependent networks robustness by adding connectivity links," *Phys. Stat. Mech. Appl.*, 444, pp. 9–19, 2016, doi: 10.1016/j.physa.2015.10.010.

70. J. Zhao, O. Yagan, and V. Gligor, "On connectivity and robustness in random intersection graphs," *IEEE Trans. Automat. Contr.*, 62(5), pp. 2121–2136, 2017, doi: 10.1109/TAC.2016.2601564.

71. A. Jamakovic and S. Uhlig, "On the relationship between the algebraic connectivity and graph's robustness to node and link failures," *2007 Next Generation Internet Networks*, pp. 96–102, 2007, doi: 10.1109/NGI.2007.371203.

72. G. S. Peng and J. Wu, "Optimal network topology for structural robustness based on natural connectivity," *Phys. A Stat. Mech. Appl.*, 443, pp. 212–220, 2016, doi: 10.1016/j.physa.2015.09.023.
73. S. Ahmadzadeh, G. Parr, and W. Zhao, "A review on communication aspects of demand response management for future 5G IoT- based smart grids," *IEEE Access*, 9, pp. 77555–77571, 2021, doi: 10.1109/ACCESS.2021.3082430.
74. B. Sliwa, R. Falkenberg, T. Liebig, N. Piatkowski, and C. Wietfeld, "Boosting vehicle-to-cloud communication by machine learning-enabled context prediction," *IEEE Trans. Intell. Transp. Syst.*, 21(8), pp. 3497–3512, 2020, doi: 10.1109/TITS.2019.2930109.
75. X. Liao *et al.*, "Cooperative ramp merging design and field implementation: A digital twin approach based on vehicle-to-cloud communication," *IEEE Trans. Intell. Transp. Syst.*, pp. 1–11, 2021, doi: 10.1109/TITS.2020.3045123.
76. A. Nair and B. R. Chandavarkar, "Temperature- and SNR-based on demand routing protocol for underwater communication," in *2021 12th International Conference on Computing Communication and Networking Technologies (ICCCNT)*, 2021, pp. 1–9. doi: 10.1109/ICCCNT51525.2021.95797982021.
77. T. Kimura and M. Ogura, "Distributed 3D deployment of aerial base stations for on-demand communication," *IEEE Trans. Wirel. Commun.*, 1276(c), pp. 1–15, 2021, doi: 10.1109/TWC.2021.3086815.
78. B. Sliwa, R. Falkenberg, T. Liebig, J. Pillmann, and C. Wietfeld, "Machine learning based context-predictive car-to-cloud communication using multi-layer connectivity maps for upcoming 5G networks," *2018 IEEE 88th Vehicular Technology Conference (VTC-Fall)*, 2018-August, 2018, doi: 10.1109/VTCFall.2018.8690856.
79. Y. Zhao, B. Li, B. Qi, and S. Tian, "Research on technologies of user side integrated demand response communication service for multi-energy coordination," *2021 IEEE 4th International Conference on Electronics Technology (ICET)*, pp. 998–1002, 2021, doi: 10.1109/ICET51757.2021.9451126.
80. Y. Wei and F. Wu, "Research on the user demand important degree of communication equipment based on QFD," *2021 IEEE 5th Advanced Information Technology, Electronic and Automation Control Conference (IAEAC)*, 2021, pp. 1067–1072, 2021, doi: 10.1109/IAEAC50856.2021.9390898.
81. S. Herrnleben, M. Pfannem, C. Krupitzer, S. Kounev, M. Segata, F. Fastnacht, and M. Nigmann, "Towards adaptive car-to-cloud communication," in *2019 IEEE International Conference on Pervasive Computing and Communications Workshops (PerCom Workshops)*, 2019, pp. 119–124, 2021.
82. C. Jayalath, J. Stephen, and P. Eugster, "Universal cross-cloud communication," *IEEE Transactions on Cloud Computing* 2(2), pp. 103–116, 2014.
83. W. Mengqiu, M. Purvis, and M. Nowostawski, "An internal agent architecture incorporating standard reasoning components and standards-based agent communication," *IEEE/WIC/ACM International Conference on Intelligent Agent Technology*, 2005, pp. 58–64, 2005, doi: 10.1109/IAT.2005.43.
84. R. Huang, W. Shi, D. Yao, C. C. Chu, R. Gadh, Y. J. Song, and Y. D. Sung, "Design and implementation of communication architecture in a distributed energy resource system using IEC 61850 standard," *Int. J. Energy Res.*, 40(5), pp. 692–701, 2016.

85. A. Danilin, S. Sawitzki, and E. Rijshouwer, "Reconfigurable cell architecture for multi-standard interleaving and deinterleaving in digital communication systems," *2008 International Conference on Field Programmable Logic and Applications*, pp. 527–530, 2008, doi: 10.1109/FPL.2008.4630000.

86. F. S. Costa *et al.*, "Fasten iiot: An open real-time platform for vertical, horizontal and end-to-end integration," *Sensors (Switzerland)*, 20(19), pp. 1–25, 2020, doi: 10.3390/s20195499.

87. F. Ye, P. Shao, Y. Guo, and J. Geng, "A data modeling method for power trading operation based on horizontal and vertical dimension integration," *2017 IEEE 2nd Information Technology, Networking, Electronic and Automation Control Conference (ITNEC)*, 2018, pp. 448–452, 2018, doi: 10.1109/ITNEC.2017.8284772.

88. L. Liu and J. Wang, "Super multi-step wind speed forecasting system with training set extension and horizontal–vertical integration neural network," *Appl. Energy*, 292(April), 2021, doi: 10.1016/j.apenergy.2021.116908.

89. V. Lukoki, L. Varela, and J. MacHado, "Simulation of vertical and horizontal integration of cyber-physical systems," *2020 7th International Conference on Control, Decision and Information Technologies (CoDIT)*, pp. 282–287, 2020, doi: 10.1109/CoDIT49905.2020.9263876.

ANNEXURE 7.1

List of reviewed papers for enhanced sensor and improved connectivity along with details of journals and publishers

S.No	Author	Journal	Year	Publisher
1	Kehl et al. [11]	HardwareX	2021	Elsevier
2	Shen et al. [13]	Automatica	2021	Elsevier
3	Jiang et al. [14]	Knowledge-Based Systems	2021	Elsevier
4	Price et al. [16]	Automation in Construction	2021	Elsevier
5	Li et al. [17]	IEEE Sensors Journal	2021	IEEE
6	Dou et al. [18]	Micro machines	2021	MDPI
7	Lin et al. [20]	Computers and Chemical Engineering	2007	Elsevier
8	Sangwan et al. [22]	Procedia CIRP	2018	Elsevier
9	Hong et al. [27]	IEEE Transactions on Instrumentation and Measurement	2021	IEEE
10	Dong. [29]	IEEE Transactions on Instrumentation and Measurement	1999	IEEE
11	Dinh et al. [30]	Computer Networks	2016	Elsevier
12	Chincoli and Liotta [31]	Sensors	2018	MDPI

13	Wen et al. [35]	IEEE Sensors Journal	2020	IEEE
14	Nazemi et al. [37]	Sensors	2019	MDPI
15	Banerjee et al. [43]	9th International Symposium on Advanced Topics in Electrical Engineering	2015	IEEE
16	Yi et al. [44]	Sensors	2018	MDPI
17	Chiu et al. [46]	IEEE International Conference on Robotics and Automation	2014	IEEE
18	Hussain and Gurkan [47]	IEEE transactions on instrumentation and measurement	2011	IEEE
19	Weddell et al. [48]	6th Annual IEEE Communications Society Conference on Sensor, Mesh and Ad Hoc Communications and Networks	2009	IEEE
20	Liu et al. [49]	International Symposium on Flexible Automation	2016	IEEE
21	Martínez et al. [50]	Sensors	2017	MDPI
22	Gomes et al. [51]	Transactions on Emerging Telecommunications Technologies	2021	Wiley
23	Bai et al. [52]	Ad Hoc Networks	2003	Elsevier
24	Mittal et al. [53]	2nd International Conference on Advances in Computing, Communication, & Automation	2016	IEEE
25	Kamya et al. [55]	Fifth World Conference on Smart Trends in Systems Security and Sustainability	2021	IEEE
26	Govindan and Mohapatra [56]	IEEE Communications Surveys & Tutorials	2011	IEEE
27	Yao et al. [57]	IEEE Transactions on Industrial Informatics	2020	IEEE
28	Rahnama et al. [61]	The International Conference on Technological Advances in Electrical, Electronics and Computer Engineering	2013	IEEE
29	Sato et al. [62]	IEEE Transactions on magnetics	2012	IEEE
30	Besnoff et al. [63]	IEEE Antennas and Wireless Propagation Letters	2017	IEEE
31	Deng et al. [64]	Applied Mathematics and Computation	2020	Elsevier
32	Banerjee et al. [65]	12th International Conference on High Performance Switching and Routing, HPSR 2011	2011	IEEE
33	Cui et al. [66]	Physica A: Statistical Mechanics and its Applications	2018	Elsevier

34	Thota et al. [67]	IEEE Network	2013	IEEE
35	Abbas et al. [68]	IEEE Transactions on Control of Network Systems	2017	IEEE
36	Ji et al. [69]	Physica A	2016	Elsevier
37	Zhao et al. [70]	IEEE Transactions on Automatic Control	2016	IEEE
38	Jamakovic et al. [71]	Next Generation Internet Networks	2007	IEEE
39	Peng and wu [72]	Physica A	2015	Elsevier
40	Kimura and Ogura [77]	IEEE Transactions on Wireless Communications,	2021	IEEE
41	Sliwa et al. [78]	IEEE 88th Vehicular Technology Conference	2018	IEEE
42	Zhao et al. [79]	IEEE 4th International Conference on Electronics Technology	2021	IEEE
43	Wei and Wu [80]	IEEE 5th Advanced Information Technology, Electronic and Automation Control Conference	2021	IEEE
44	Herrnleben et al. [81]	IEEE International Conference on Pervasive Computing and Communications Workshops	2019	IEEE
45	Jayalath et al. [82]	IEEE Transactions on Cloud Computing	2014	IEEE
46	Costa et al. [86]	Sensors	2020	MDPI

Chapter 8

Reconfigurable robotic systems for Industry 4.0

Ekta Singla

8.1 INTRODUCTION

Industries have evolved from mechanization to the inventions of electrical energy and strengthening through the times of automated production systems towards the era of Industry 4.0. Constituted by the advanced technologies, say the Internet of Things (IoT), cloud-based analysis, data analytics, smart manufacturing systems, virtual and augmented reality, etc., the current industrial revolution possesses multifold facets. Amidst the broad spectrum of this cyber-physical system, the adaptivity of physical systems plays an important role!

The physical systems include the machines for various manufacturing processes, material handling systems, robotic systems, and all such devices which are connected with sensors and communication devices and are associated with the industrial domain. The advanced digital platforms do connect these systems together through cyber networking, for improved productivity, predictive analysis, and optimized energy consumption; but such cyber-physical systems can be effective only when the corresponding physical systems can respond to the information. This chapter focuses on the importance of adaptivity of the industrial robots to the variations in the environments, and details the modular and reconfigurable robotic systems.

8.2 CUSTOMIZED ROBOTIC MANIPULATORS

A robotic system happens to be an integral part of industrial automation and the diversity in such environments demands specialized designs of robotic manipulators and their work cells. Robotic manipulators, the arm-like mechanisms consisting of a series of links usually connected with revolute or prismatic joints, play important roles in industrial automation. The conventional manipulators may cover a significant range of tasks, say material handling, welding, room cleaning, and operations in hazardous environments, inspection, maintenance or assembly; however, with the rapidly growing range of less repetitive applications – the concept of customized

DOI: 10.1201/9781003246466-8

robotic manipulators is emerging as the need of the day. The underlying problem is the necessity of layout design in accordance with the installation and working of these standard manipulators. A lot of human effort and time is spent to plan the required changes in the layout. This can be considered workable for a few specific cases but may be inappropriate when we plan to use the manipulators at several places and for a variety of applications. Instead of this, we can look at the problem with a different perspective: "Given a workcell, can we work out a procedure to design a robotic manipulator which is able to work efficiently at the desired locations of this environment, while avoiding the obstacles?" In general, to design and fabricate an altogether new manipulator to accomplish an assignment may not sound cost-effective. However, in many situations, reconstruction of an existing layout so as to accommodate standard robotic arms, is certainly not desirable. Industry 4.0 involves timely data collection, timely status information, history, and target states, which help in optimal planning of production systems. Demand-based customized products can be planned for best utilization of resources and services. This involves highly flexible mass production that can be rapidly adapted to market changes. Work pieces, tools machines, and robots need to be capable of autonomously exchanging information, triggering actions, and controlling each other independently.

New trends towards mass customization, small scale manufacturing, and maintenance services – which are non-repetitive in nature – require robotic manipulation systems with different configurations. Customized robot configurations are desired on a regular basis, to cope with the frequent changes in task and environment. To provide cost-effective solutions for rapid customization, modularity and reconfigurability have provided promising solutions.

8.3 RECONFIGURABLE MANUFACTURING SYSTEMS: A BROAD PERSPECTIVE

To address the issues associated with flexibility and fluctuations in customer requirements, reconfigurable manufacturing systems (RMS) are receiving huge attention [1–3]. They involve a rapid change in hardware and software components. Some of the important characteristics of a reconfigurable system include flexibility, modularity, convertibility, and scalability [4]. These terms are discussed ahead from the utilization perspective, and how to represent how modularity has taken a keen position in the Industry 4.0 era. Reconfigurable robotic systems are gaining attention, so as to provide options for being adaptable to the fluctuations in demands. Conventional systems possess limited flexibility, which can be attained through reprogrammability of robotic systems, and also through reorientation of flexible manufacturing systems (FMS). Keeping the goal to quickly change the production system, as a solution to fluctuations in market demand and for handling market turbulence, some of the prominent characteristics of RMS are discussed here.

o *Modularity:* To cater to a vast set of workcell combinations, design and planning of modular workcells and modular sub-systems come out to be effective in forming reconfigurable combinations. Modularity is an inherent aspect of an RMS and several combinations/assemblies can be planned for reconfiguring a manufacturing system as desired.

o *Integrability:* For well-planned modular sub-systems – ready to be assembled or arranged for developing an RMS – an essential requirement is the integrability among the modular sub-systems. Since the major key features of Industry 4.0 are minimizing down-time, safety, and maximum utilization of resources – fast reconfiguration and integration possibility is an important feature.

o *Scalability:* To plan a manufacturing system adaptable to variations in market fluctuations, the expansion and scaling down of the working system should be possible. This feature of a manufacturing system to be capable of adapting the production scale in a cost-effective manner is called scalability. This can be referred to through the examples of expanding material handling capability through changeable size of conveyer systems, adding numbers of mobile robots, addition in carousal storage systems, etc.

o *Diagnosability:* While adapting reconfigurability in a system, it becomes important to rapidly diagnose the ramp-up/down of the new manufacturing configuration – both economically and in production efficiency. Well-designed performance measures are required to analyze the possible configurations. Majorly, optimal configurations are designed a priori before the change of the set-up physically. The efficiency, production rate, inventory measures, etc., are the normally used performance indices.

o *Universality:* One of the major features associated with the new industrial trends is their applicability universally. A global aspect is important for any new system to be utilized in a manufacturing system, be it related to safety measures, efficiency, and/or system layouts. The modular sub-systems are planned to be global – both in technical and utilization capability.

o *Compatibility:* Similar to integrability, the modular components, i.e., the sub-systems, should possess compatibility among themselves. This is related to the sizes, locations, orientations, speeds, pressure, carrying capacities, loading-unloading strategy, system resolutions, etc.

8.4 MODULAR ROBOTIC SYSTEMS

Robotic systems are an inherent part of the reconfigurable manufacturing systems, and the concept of modularity is detailed through the discussion on these important components of RMS. The primary advantage of modularity is associated with re-configurability. It signifies the possibility of on-site variation of the configuration of a modular manipulator for different task-based requirements. Reduction in repair time and fault tolerance are the advantages of modular developments, and these are prime requirements of advanced automation.

In reported works, modularity has been utilized for various industrial serial manipulators, in applications of pick and place, milling operations, and assembly automation [5–7]. Some of the reported works have focused on the scientific facilities such as cluttered environments of power-plants, etc., for discussing the role of modularity and reconfigurability [8-10]. A common aspect of all these attempts is that the modelling of the system is worked upon after the reconfiguration of the built hardware. This needs several attempts to obtain the configuration most useful for the required task. To control the system, its kinematic and dynamic model is developed through measuring the geometrical and inertial features of the configured system. Some of the works focused on avoiding the hit-and-trial system and to develop systematic design strategies using modular architectures. There are two ways reported to proceed in that direction. Either by computing the robotic parameters which can be adapted by the modules for the required assembly, or by designing and developing the modules in such a way that the design programs can directly give the optimal combinations for a given set of tasks.

The design of a modular manipulator, links architecture, divisions, and assembly planning of the modules are briefed in Figure 8.1. Also, a broad

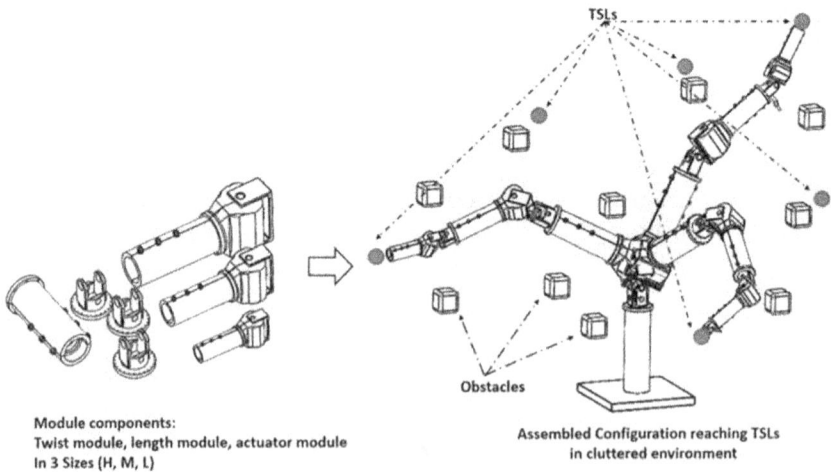

Module components:
Twist module, length module, actuator module
In 3 Sizes (H, M, L)

Assembled Configuration reaching TSLs
in cluttered environment

Figure 8.1 An example of a modular reconfigurable serial manipulator, designed for working inside a cluttered environment for required working locations

perspective of types of applications which can be catered to through one reported set of modules is included here, for clarity.

8.4.1 Design challenges

The objective of a modular system design is to provide a solution for a given set of applications. The designed size, type, shape, weight, and other capabilities of modular units are expected to be combined together as per the requirements and get controlled. There are several challenges which need attention while planning the design. The foremost issue is related to the number and type of modules. Can one active modular unit have the ability to carry several units of the same kind? If so, what will be the limit of such a number? Will the solution be power and cost effective? If types are to be designed then what are the appropriate criteria to decide so? How many types of modular links or joints will be sufficient? In all, how can the working of the assembled modules be ensured? There are major discussions related to modelling and control techniques! Further sections give some insight into these aspects.

8.4.2 Basic classifications of modular links

The modular links and joints are basically classified as active or passive types, homogenous or heterogenous types, reconfigurable or self-reconfigurable types. Here, the actuated links are referred to as active links which are controlled for performing the required motion. There can be some extensions or extra units which contribute passively in extending the link-lengths, providing balance or for adding flexibility. The passive links and joints do add extra features but need attention in controller planning. In the second type of classification, if all the modules are of the same kind then those are referred to as homogeneous links and joints, but if the units are different in features in any form then those are referred to as heterogeneous modular components. The latter deals with a large variation in types of components and that is not desirable in the modular robotic systems. In general, hybrid systems are recommended where variation is planned optimally. In the third type of classification, the modular assemblies are referred to as reconfigurable if their configuration can be changed without disassembling the system. In a case where the reconfiguration can be acquired without any manual assistance, then the system is referred to as self-reconfigurable.

8.4.3 Module architecture

A module is a unit or a set of units which is connected to other modules of the same or different types to develop an assembly of required features. Figure 8.2 may be referred to for exemplary designs. While planning the

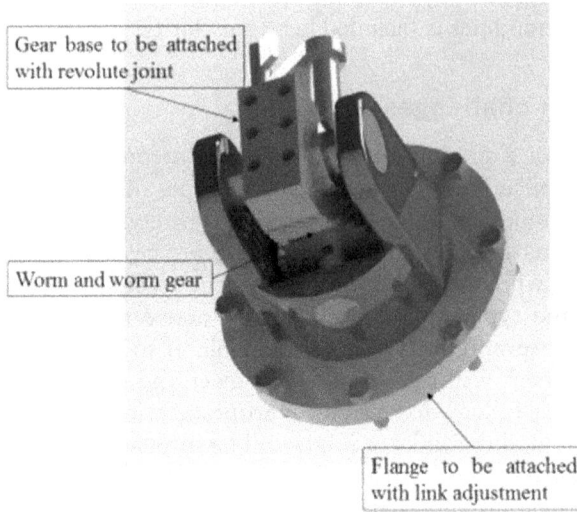

Gear base to be attached
with revolute joint

Worm and worm gear

Flange to be attached
with link adjustment

Figure 8.2 Exemplary modular units of a robotic system. (Courtesy: MOIR lab, IIT Ropar)

internal architecture of a module, the following are the foremost aspects to be decided upon.

1. *Link-length unit:* Normally, the link length unit is a passive unit which provides length to a module. The units are planned in different lengths. In some works, the unit is an inherent part of the joint module with a fixed link-length. For change in length, extensions are planned. These extensions are kept in less variations, as any change in a parameter leads to change in the kinematic and dynamic models. A corresponding description is detailed ahead.

2. *Joint unit:* This unit consists of a casing with transmission system actuator, encoder, and required torque sensors, etc., and connected to length unit. The joint modules are normally active units, but sometimes to add on flexibility some passive joints can be planned. The architecture of these active-passive modular units is differently planned in various reported works.

3. *Links-connectivity adjustment unit:* This is a small part between a basic link-length module and the next link, for the adjustment in connectivity angles, if required. In conventional systems – which include almost all industrial robotic configurations – these angles are multiples of 90. Many of the works have not given this provision through adjustment, but have added fixed modular units for parallel connection or perpendicular connection.

8.4.4 Assembly planning

The number of degrees of freedom can vary within prescribed limits. The number and the locations of working points, payload capacity, obstacles avoidance, and other such task-requirements can be considered as the required inputs before designing a reconfigurable robotic system. The number of degrees of freedom is decided through optimization techniques. These bounds are decided based upon the modular architecture.

Which modules can be used as base modules? How many modules of one kind can be carried by their own and by others? Is there any specific module to be added in distal end, etc.? All such decisive points are compiled in the form of assembly rules, which can help the user while planning the assemblies

8.4.5 Multi-layer approach to modularity

A multi-layer approach showcasing the spectrum of possible service applications of modular robotic systems is presented in Figure 8.3. It proposes a layered approach which is similar to group technology. Similar features and required tasks are clubbed together, and specifications of an applicable modular library can be defined. This ensures that customized reconfigurable systems can be made available for a larger set of applications. It is highly unlikely to have a common solution for all required robotic applications, as the required reachability, payload capacity, weight intensity, workspace, etc. are different. A trade-off is required between the maximum variations

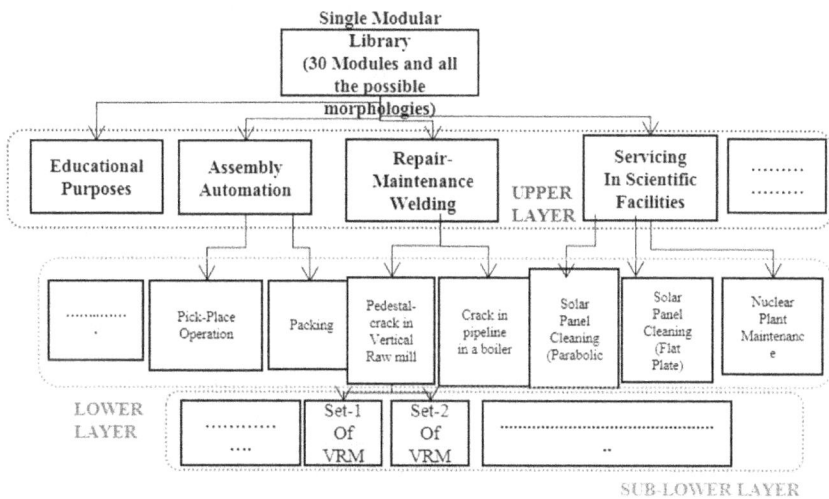

Figure 8.3 A multi-layer approach to showcase the spectrum of possible service applications of modular robotic manipulators

which can be catered to using the same library, and the minimum efforts/ torque requirement in running the systems.

Figure 8.3 presents a multi-layer approach to showcase a broad spectrum of applications of modular libraries. The broader differences lead to the separation of industries at the upper-layer. Those are quite different in requirements from each other. As the layer decreases – the similarities can be observed. These are from the perspective of robotic assistance and robotic specifications – as their size, degree of freedom, torque requirement, etc.

On further dissection, the next layer of classification, i.e., lower layer, divides the modular library in its corresponding set broadly, on the basis of different environments and arrays of manipulator-morphologies used for each of these environments. Each of these environments, along with its procedural-operations and array of morphologies required to implement them in different workspaces, constitutes a single job which is implemented at one time. Here, in the first environment more redundancy is required than in the latter as it is more cluttered, and welding procedures can be different in both. This justifies the classification of the lower layer. Similarly, in the case of servicing in scientific facilities, different environments can be solar plant and nuclear plant, which may have the same field of upper layer – say welding, but at lower layer the requirements may vary – that leads to the application area of modular reconfiguration systems.

8.5 WORKSPACE RECONSTRUCTION FRAMEWORK

The utilization of modular and reconfigurable design approaches is worked with the information of environment models and working locations. It is important for advanced systems to reduce the human intervention in taking measurements, and rather scanning technologies can be utilized for automatic workspace model constructions. Workspace is captured through one or a set of three-dimensional depth images of the workspace, and these are recorded in the form of point clouds. The entire workspace may be too big to be recorded in a single image. The distribution of the point clouds in the image represents the geometry and the clutter of the workspace. The orientation and position of the cameras can be recorded through inertial measurement units. The information from these units is transformed using designed matrix generators. The output can be represented simply as the transformation matrices from point cloud frames to the reference frame. Later, the recorded points are assembled to form a single one with respect to the standard frame, in point cloud data (PCD) format. In the second phase, these point clouds are further processed to downsize the total number and get into surface reconstruction. This is challenging. Even though there are many widely used algorithms, constructing reliable and accurate models from point clouds remains an open problem. It gives good performance even when the data is noisy and of varying density. It also gives near real-time performance. This step takes in the *PCD* file as the input and generates an *STL* file as the output. A mesh representation is necessary for running the optimization processes in the later part of the framework.

REFERENCES

1. Bortolini, M., Galizia, F.G., Mora, C. (2018). Reconfigurable manufacturing systems: Literature review and research trend. *Journal of Manufacturing Systems* 49: 93–106.
2. Fusko, M.I., Rakyta, M.I., Krajcovic, M.A., Gaso, M., Grznar, P. (2018). Basics of designing maintenance processes in industry 4.0. *MM Science Journal* 2018(March). 2252–2259 DOI: 10.17973/MMSJ.2018_03_2017104
3. Maldonado-Ramirez, A., Lopez-Juarez, I., Rios-Cabrera, R. (2019). Reconfigurable distributed controller for welding and assembly robotic systems: Issues and experiments. In Shanben Chen, Yuming Zhang and Zhili Feng (Eds.) *Transactions on Intelligent Welding Manufacturing* (pp. 29–49). Springer, Singapore.
4. Mourtzis, D., Fotia, S., Boli, N., Vlachou, E. (2019). Modelling and quantification of industry 4.0 manufacturing complexity based on information theory: A robotics case study. *International Journal of Production Research*. 10.1080/00207543.2019.1571686
5. Acaccia, G., Bruzzone, L., Razzoli, R. (2008). A modular robotic system for industrial applications. *Assembly Automation* 28(2): 151–162.
6. Chen, I.M. (2015). Modular robots. In Nee, A. (eds) *Handbook of Manufacturing Engineering and Technology* (pp. 2129–2168). Springer, London. https://doi.org/10.1007/978-1-4471-4670-4_100
7. Song, L., Yang, S. (2011). Research on modular design of perpendicular jointed industrial robots. In: Jeschke, S., Liu, H., Schilberg, D. (eds) *Intelligent Robotics and Applications* (pp. 63–72). Springer, Berlin Heidelberg https://doi.org/10.1007/978-3-642-25486-4_7.
8. Aghili, F., Parsa, K. (2007, August). A robot with adjustable D-H parameters. In F. Aghili and K. Parsa, *International Conference on Mechatronics and Automation*, Harbin, China, 2007, pp. 13–19, doi: 10.1109/ICMA.2007.4303509.
9. Pagala, P., Ferre, M., Armada, M. (2014, January). Design of modular robot system for maintenance tasks in hazardous facilities and environments. In Armada, M., Sanfeliu, A., Ferre, M. (Eds.) ROBOT2013: First Iberian Robotics Conference. Advances in Intelligent Systems and Computing, vol 253. Springer, Cham. https://doi.org/10.1007/978-3-319-03653-3_15
10. Singla, E., Tripathi, S., Rakesh, V., Dasgupta, B. (2010). Dimensional synthesis of kinematically redundant serial manipulators for cluttered environments. *Robotics and Autonomous Systems* 2010.Volume 58, Issue 5, Pages 585–595, ISSN 0921-8890, https://doi.org/10.1016/j.robot.2009.12.005.

Chapter 9

Introduction to additive manufacturing

Concepts, challenges, and future scope

Sumitkumar Rathor, Ravi Kant, and Ekta Singla

9.1 INTRODUCTION

The fourth industrial revolution (Industry 4.0) includes robotics, artificial intelligence, Internet of Things (IoT), big data, and additive manufacturing. The ongoing Industry 4.0 has nine pillars, shown in Figure 9.1, which provide many opportunities for research and applications [1, 2]. It is the generation of computers, mobiles, and other digital devices. Cybersecurity is vital to keep the data and information safe and secure. Augmented reality provides tools to interact, connect, and experience the virtual world. Many products can be digitally visualized and experienced by enhanced virtual

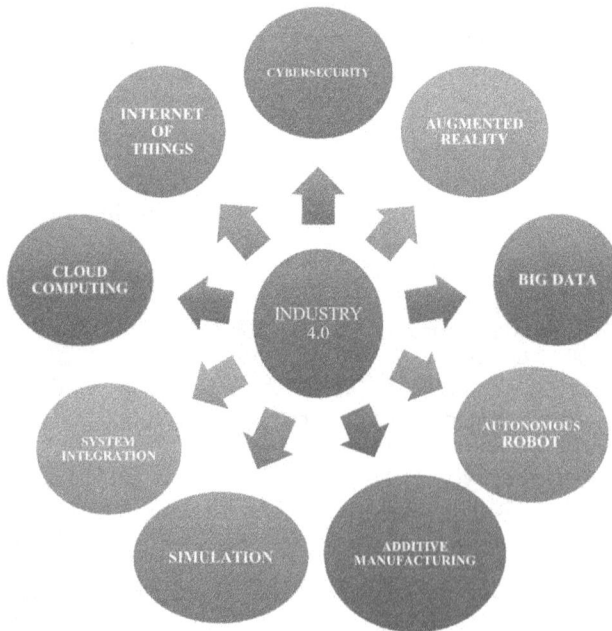

Figure 9.1 The nine pillars of Industry 4.0

DOI: 10.1201/9781003246466-9

versions. All the generated digital data is managed systematically to analyze and process. Before 3D printing, big data can improve parameter selection, geometry validation, and similar characteristics decisions. It will enhance the process planning and process chain for mass parts production. Industry 4.0 changes the face of the current industry while implementing digitization in process and work. Intelligent machines are getting popular in the routine operation of the manufacturing industry for achieving improved speed and accuracy. Digital data and intelligent machines will work for and from all the areas of the globe.

Simulations create the engineered, optimized design data files. The raw digital files are used with different simulation tools like Ansys and Simufact. This software offers optimized design for additive manufacturing (DfAM) to understand and visualize the complex phenomenon of mechanical and thermal behavior of the part. Whenever data cannot be carried or stored in an ordinary system, some Internet-based remote servers' infrastructure is accessed, called cloud computing. An application-based software provides seamless connectivity to linked machines. The Internet of Things is a network of physical objects embedded with software, cloud, sensors, and networks. It is now being implemented in appliances, autonomous farming equipment, wearable health monitoring, and intelligent factory equipment. The Industry 4.0 work environment is adapted when additive manufacturing techniques are flexible enough. The seven major AM techniques are as follows:

1. Vat polymerization.
2. Material jetting.
3. Binder jetting.
4. Material extrusion.
5. Powder based fusion.
6. Sheet lamination.
7. Direct energy deposition.

The AM has limitations in printing time, surface roughness, and reproducibility for these simulations and modeling that will be implemented to improve the AM process in the future [3].

Evolving digitization can create new AM technologies to overcome current design-related hurdles. With the improvements in cyber technologies, designers can use computational facilities to enhance efficiency and productivity. Design for additive manufacturing (DfAM) is a new design tool introduced recently for process parameter selection (quality, time, cost, reliability, and CAD constraints). The methodology is to design new products or existing ones, in which customer requirements, functions, design parameters, and process variable parameters are evaluated simultaneously [4]. We see the applications of AM technologies as a pillar of Industry 4.0

in different areas like healthcare, marine, aerospace, maintenance, etc. [5]. Scientists and researchers are working on technologies and methods which will have more impact on human life. With improvements in sustainability in AM technologies, energy consumption is considered a critical approach. The new methodologies of life cycle energy analysis and greenhouse emission saving are also considered in this new manufacturing era [6]. AM can produce a part of a product with minimum human involvement in less time. The COVID-19 pandemic introduced the need for customized products with complex shapes, less weight, and that are available in less time [6]. AM has played a significant role in fulfilling the need for that time as it has the ability to tackle variation in design, delivery time, and volume of production. The discussion on basic concepts, challenges, and future applications involved in AM technologies for enhancing the competencies of IoT is discussed in this chapter.

9.2 ADDITIVE MANUFACTURING

Additive manufacturing is the regularized word for rapid prototyping (RP), also known as 3D printing. The word prototyping is used to describe the process of generating a model for a representation or something which will be created in less time in the industry. These prototype products can be used for establishing a rigid manufacturing process that is ready to go in the market. The ideas of someone will be converted into a physical product in less time using RP. The AM uses computer-aided design (CAD), or 3D object scanned object files to fabricate directly without many resources and planning [7]. The geometric shapes manufactured using AM have precise dimensions if there are correctly selected parameters. The traditional manufacturing processes use complex process planning, tooling, machining, and surplus material to remove during machining, which will produce more wastage of time, money, and resources. The art of AM works by adding material layer by layer, and each layer is defined by CAD file data. As the layer height reduces, the corresponding part is close to the dimensions of the final part. Many AM machines are available in the market that generally use a layer-based technique.

9.3 THE GENERAL AM PROCESS CHAIN

AM implies several steps as shown in Figure 9.2 to fabricate a physical component using CAD data. As discussed, AM needs less tooling and machining for post-processing of parts to clean or finish the rough surfaces. The following are some common steps followed in the AM process.

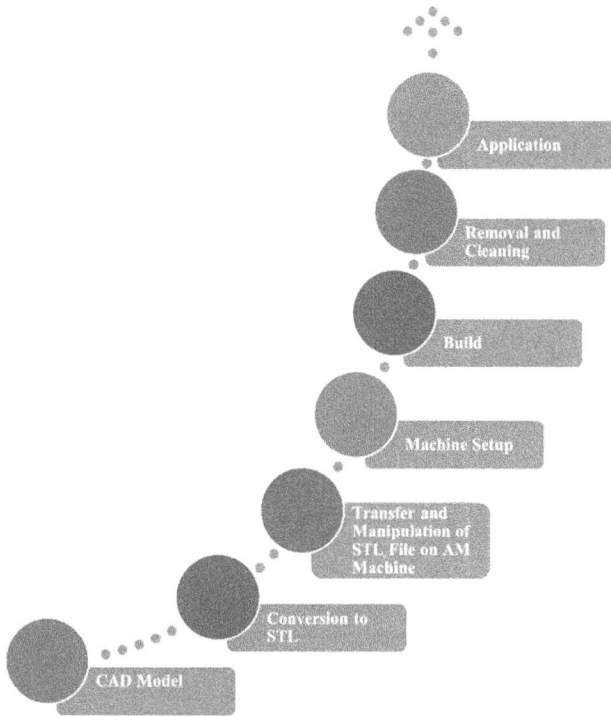

Figure 9.2 General AM process chain

9.3.1 CAD model

The first step is to create a 3D CAD model as per the geometry required for the print. The CAD model is designed using computer-aided design (CAD) software or reverse engineering techniques using an object laser scanner.

9.3.2 Conversion to STL

The CAD files are converted to STL (standard tessellation language) file format. The conversion to an STL file includes the tessellation of shapes, and the time needed to do this depends on computer specification and model complexity. Here the STL file is digitally sliced into layers.

9.3.3 Transfer and manipulation of STL file to AM machine

The STL file is transferred to the AM machine with some manipulations like orientation, printing position, and size. This leads to printing economically by cost savings and reducing material waste.

9.3.4 Machine setup

The setup parameters, such as layer thickness, print time, melting temperature, etc., are specified before printing on the machine, as most of the AM machines are equipped with such a facility. The parameter selection is important to print a part with good resolution and in less time.

9.3.5 Build

The building process of a part is carried out fully automatic without human intervention. The machine is controlled by a computer for printing parts layer by layer. The size of the part depends on the machine size.

9.3.6 Removal and cleaning

The part is removed from the machine after the printing process. The support structures are generally separated from the main part after removal. Here all the cleaning and removal of parts is done manually.

9.3.7 Post processing

The printed part is processed as per the application requirement in this step. The polishing process, heat treatment, coatings, etc., are included in the post-processing. Different applications require a high or moderate surface finish, which results in the time needed for post-processing. The quality of the building part depends on the machine and its parameters, which decide the method and time required for post-processing.

9.3.8 Application

After post-processing, the part is ready for use. There are many applications for which assembly, painting, graphics, etc., are required for more effective working. More working features are added to the part to improve the performance.

9.4 ADDITIVE MANUFACTURING TECHNOLOGIES

9.4.1 Vat polymerization

Vat polymerization is the first kind of AM technology introduced in the 1970s with a patent. As shown in Figure 9.3, the process creates 3D objects by local curing of liquid resins through a light-activated polymerization. Stereolithography is the first AM process commercially available in the vat polymerization technique. The equipment contains a vat, a light system, and an elevated platform. The process of curing a resin by shining light on

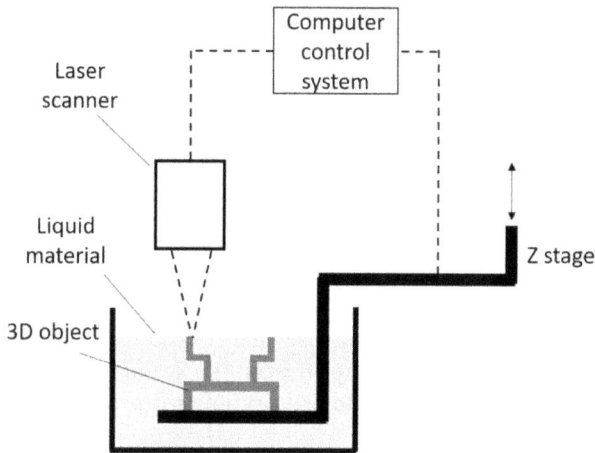

Figure 9.3 Scheme of stereolithography [8]

it is called photopolymerization. Most of the polymer reacts to the radiation in the ultraviolet wavelength, and the visual system is also used for some polymers. Upon irradiation, the photoinitiator reacts with polymer and makes polymer chains. After this chemical reaction, the material becomes solid. An exemplary laser beam is guided by scanning galvanometers to radiate over the resin surface. The resin surface traced by the laser will be cured to make 3D objects.

Laser scanning is used at a straight velocity on the surface of the resin. It has three key assumptions considering constant photopolymer. First, the laser irradiance distribution is Gaussian, a normal distribution that is symmetric about the mean, and in the graph, it will appear as a bell curve. Second, the photopolymer resin obeys the beer-Lambert law of exponential absorption. The third is the transition of the resin from the liquid phase to the solid phase at the "gel point," a sudden change in the viscosity of a polymer solution leading to a gel formation.

The step-by-step process:

1. The build platform is lowered one layer down from the top of the resin.
2. A light is shined over the surface of the resin to cure. The platform continuously moves down till the last layer is completed.
3. A blade is moved in between layers to provide a smooth resin base to build the next layer.
4. After completing the process, the vat is drained of the resin, or the elevated platform is moved upward to remove the object from the platform.

5. Significantly more minor post-processing of objects is required after the stereolithography process.

The layer thickness is constrained by the wavelength of the laser and spectral position to limit the penetration. The curing depth is calculated by using the following equation [9]:

$$C_d = D_p ln ln\left(\frac{E_{max}}{E_C}\right) \tag{9.1}$$

Where D_p is penetration depth at beam intensity reduced to 1/e of the surface value, E_{max} is maximum exposure energy, and E_c is the critical energy required for the transition of the resin from the liquid to the solid phase. As this process is introduced for rapid prototyping for product demonstration, it now gets an extension to produce functional parts. The challenges include the introduction of new materials. The development and polymerization of curing mechanisms, light source, and wavelength selection need to be adaptive. Nanoparticles and other nanomaterials are introduced with this process. The need is to exhibit the required mechanical, magnetic, and electric properties. This includes graphene, carbon nanotubes, metal nanoparticles, and ceramics [10–13]. The graphene composites show a good combination of increased strength and increased ductility. The metal nanoparticles can be used in commercially available photocurable resins to enhance their electrical properties. The surface roughness, microstructure, and mechanical strength can be enhanced by adding carbon nanotubes into the photocurable resin. The polymerization process also can be improved by adding ceramic material into a photocurable resin.

Vat polymerization is a widely accepted AM process in different application areas. This process considers many process characteristics to achieve accuracy and accessibility [14]. Recent developments in 4D printing emphasize on use of new materials to transform the shape and size of the building part. Materials are a more important factor in this process than the machine. 4D printing extends the existing 3D AM technology where built objects are transformed into preprogrammed structures after introducing external stimuli such as humidity, light, or heat [15]. Researchers are investigating novel solvent-based slurry stereolithography for the fabrication of porous ceramic membranes used for filter applications [16]. Ceramic membranes are characterized for thickness and surface roughness to achieve better efficiency. An experimental study of the compression behavior of MIP-SL material is presented by [17]. They show there are layer and connection effects between manufacturing layers. Very few mechanical studies of the stereolithography parts are becoming constraints to a wide range of applications. Effects of build orientation on polymers' mechanical properties and behavior are investigated [18]. They show that layout significantly affects tensile and impact properties, but the axis of printing does not affect much. Similarly, strain rate, layer thickness, and size effects on Young's Modulus, ultimate tensile strength, and fracture strains are investigated [19]. Anisotropy in mechanical properties is observed in different planes and showed mechanical property dependency in the built part. AM denture stress distribution is evaluated in different print angle directions (0^0, 45^0, and 90^0) [20]. The stress distributions of 3D printed dentures are observed, and the maximum principal strains are less in 45° printing directions. It is an active area of research for a wide range of industrial applications.

9.4.2 Material jetting

Material jetting is the process patented by Object Ltd. In 1999 they used the name PolyJet. This is the process that combines Inkjet technology with the use of photopolymers, shown in Figure 9.4. The most common materials used for material jetting are casting wax and photopolymers. Material jetting is one of the quickest and most accurate processes for AM. The liquid photopolymer droplets are cured to form a new layer with ultraviolet light. All other remaining layers are built on top of the previous layer, and in some cases, layers are allowed to harden and cool. An ultra-thin layer of photopolymer layer material is built on a build tray, and direct ultraviolet light is shined following the same nozzle travel track.

Figure 9.4 shows the use of ceramic material ink to build the green and sintered part by the material jetting process. This technique allows the printing of the layer size in microns with nearly fully sintered parts. The dimensions of the parts sintered are more accurate, but the other characteristics, like mechanical strength and roughness, are anisotropic. It is more along the build direction than different directions, mainly because of the support structures. The key challenging aspect is the improvement in obtaining bonding strength between the layers. Wettability between the deposited top and bottom layers can also be crucial to producing parts

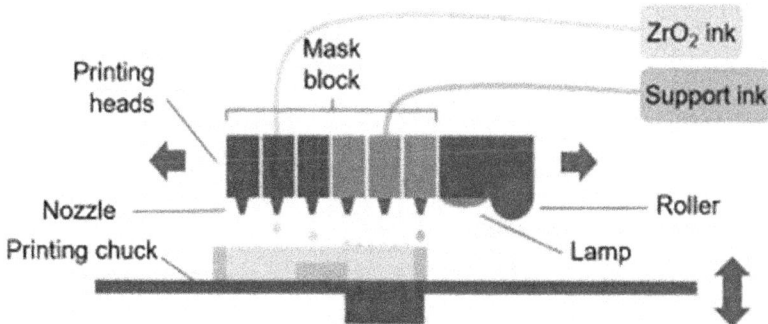

Figure 9.4 Scheme of material jetting [21]

with good mechanical strength. Build orientation selection is important in designing the part for optimum use of support structures. Formulating the liquid material is a complex procedure and is not easy. The materials available in other than liquid form, the material particles, are needed to be used with some dissolvers, solvents, or melting of the material if possible. After this liquid formation, it is necessary to convert the liquid into discrete droplets. It can depend on the relationship between machine capability and process parameters. The control of the amount of deposition of these droplets is a primary function of the print head, process parameters, and the travel trajectory of the printing nozzle. It is important to control droplet deposition to achieve the surface finish and to build accurate shape parts.

For the material jetting process, very few materials are available for processing. Mainly waxes and polymers are the commonly used materials as they have the capability to form drops due to their viscous nature. Industries are developing machines and improving materials that can be processed using this technology. The recent most studied materials include polymers, ceramics, and metals. As low viscosity plays a role in forming droplets, the viscosity of the materials is lowered. Polymer materials give a range of mechanical properties and applications. Deposition modes of materials are important as this relates to the deposition speed of the process. Functionally graded materials mainly containing composite base materials ABS-like and rubber-like have good mechanical properties [22]. Ceramic materials are less corrosive, heat-resistant, hard, and brittle material. Ceramic powder is used with solvent and other additives to form a required viscous liquid for processing. There may be chances of uneven deposition of the material layer by layer, so larger shape parts are challenging to build. Also, the part printed with this technology will be dense, and shrinkage may affect the dimensions of the parts. Metal powders of copper, aluminum, and tin have been used to build the part. The melting point of the metal is high, so while processing, it may damage the parts of the machine. The design of the nozzle is important for the metal powder solution to form droplets of the metal solution. The microstructure and porosity of the built part are very impressive.

The application parts built by material jetting technology include robotic mechanisms [23], special prosthesis [24], custom tooling [25], and multi-material manageable mechanisms [26]. Contactless and in-situ monitoring of ceramic manufacturing is ideally suited to optical methods. Laser speckle photometry (LSP) is the non-destructive technique for investigation of the process stability material and quality in material jetting [19]. Investigation of mechanical properties and understanding of processing limits of zirconia with different conditions are identified [19]. The orientation of specimens printed relatively shows higher strength in 0^0 orientation under compression and flexural conditions. The effects of process parameters on the mechanical behavior of parts fabricated by multi-material jetting are investigated

[27]. The study aims to characterize the effects of process parameters on the mechanical properties of VeroWhitePlus and VeroClear test specimens. VeroClear specimens have more impressive results when compared to the mechanical properties of VeroWhitePlus specimens.

9.4.3 Binder jetting

The binder jetting technology uses binder material with the base material to form a 3D part. The binder is deposited into the powder bed to build objects. The process, shown in Figure 9.5, has a powder bed with a nozzle to deposit the binder material using three-axis movement. The binder droplets form a spherical agglomeration of powder and binder liquid. A small amount of material is jetted through a nozzle. The print head is moved in the horizontal direction to deposit an alternating layer of the binder material. Material is spread on the powder bed, and the binder is jetted following part geometry after every layer. This technology is different from the material jetting technology, where part material is jetted from a print head. Here, binder or other additives are deposited on a powder bed through the print head [28]. After completing the process, the printed part is known as the green part. The post-processing is carried out to clean the part after removing it from the bed and make the part stronger by infiltrating the part. Support structures need very little as the parts are self-supported in the powder bed [29]. Prototyping is the first application of this technology as some applications can also be seen for production purposes. A specifically designed material composition can obtain lightweight parts with function prototypes.

A wide range of materials is available for binder jetting technology. The different powder mixtures of composition and size variation are being

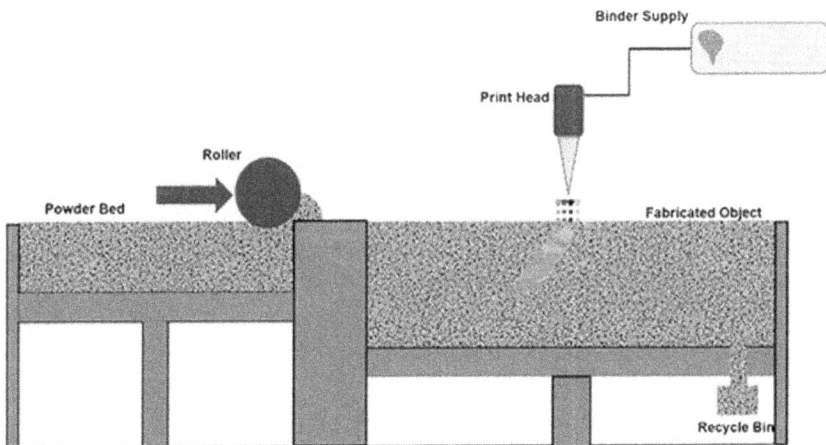

Figure 9.5 Scheme of binder jetting [30]

investigated. Industrial applications include materials like silica sand, 316 stainless steel, silicon carbide, ceramic, and IN625. Creating complex geometry structures with precision and accuracy makes this process common for sand mold and cast pattern use. Also, this process is relatively quick and affordable compared to other AM technologies. Binder jetting allows for the making of full-color prototypes and low-cost and high-volume production. The efficiency of powder packing and wettability with a binder in the green part is important as it otherwise restricts the final product's densifications and strength.

Mechanical properties, microstructure, apparent densities, and porosities of sintered parts are investigated with different mixtures of alumina [31]. A few research works focus on individual aspects of binder jetting, including material evolution under different processing conditions of sintered parts. This technology is investigated for sensitive alloys as there is no need for melting and solidification of material to build the part. The effects of different process parameters like thermal, mechanical, and porosity are explored for binder jetted 316L samples [32]. This study shows the potential of using this technology for different alloys in the near future. In the binder jetting process, the analysis of droplet size and its impact with penetration into powder bed can be a challenging area of research. The capillary flow in tubes of different sizes is investigated to understand the liquid flow in a porous material [33]. They proposed binder droplet interaction with the powder bed and matched the numerical results with experimental ones.

9.4.4 Material extrusion

Material extrusion technology is one of the very popular AM techniques adopted by industries and academia for different purposes. This technology, shown in Figure 9.6, uses a continuous filament made of thermoplastic material to build the part. There is a feeding system from the coil to the nozzle that the filament is fed from. The print head contains a heating coil, wire feed mechanism, and nozzle. Material is passed through the nozzle after heating and deposited layer by layer. The nozzle can move horizontally, and the platform can move up and down vertically to deposit material after each completion of a layer. This process builds parts layer by layer, and constant pressure is applied through the nozzle to get a continuous flow of material. The quality of the finish, accuracy, and speed depend on nozzle geometry as the perfect square nozzle is difficult to make, so it always has a radius [34]. Similarly, gravity and surface tension are important factors to be considered for the high tolerance required parts. Considering simple screw geometry, the heated material flows through channels of the screw to the nozzle. So, the velocity w of material flow is calculated by

$$w = \pi DNcoscos\,\phi \qquad (9.2)$$

A)

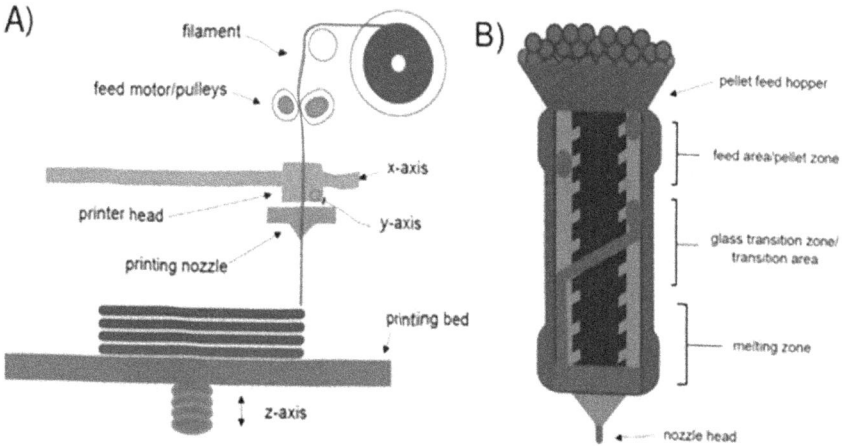

B)

Figure 9.6 Scheme of (A) fused filament fabrication (B) single screw extrusion printer head [35]

where D is the diameter of the screw, ϕ is the screw angle, and N is the screw speed. After extrusion of the material from a nozzle, it should hold a similar size and shape. Surface tension and gravity can change the shape, and cooling can affect the size of the extruded form of material.

Due to zero velocity of the melt at nozzle boundaries, it sticks to the nozzle walls. This material is subjected to shear deformation during flow. Therefore, shear stress is defined as

$$\tau = \left(\frac{\gamma}{\varnothing}\right)^{\frac{1}{m}} \tag{9.3}$$

where m is the flow exponent, γ is the shear rate, and $\varnothing$ is fluidity. The support structures are introduced to support the main geometry with point contacts. The overhang features and other low-angle features are supported with support structures. The support structures can consist of the same material or some different material of the same category. A separate extrusion chamber is used for the generation of supports of a different kind. It is very important to design supports carefully. It is supported and can be used to increase the heat transfer from the building geometry by adding more surface area. Support structures can be removed easily; otherwise, they will damage the part's features and characteristics. These are needed to be comparatively weaker than the parent part so they can be removed with the use of light tools. Popularly used materials are ABS, nylon, and polycaprolactone. Tensile strength and high-density parts are expected from these materials after using material extrusion technology or the fused deposition method (FDM) [36]. Generally, properties are isotropic in the x-y plane, and some design considerations need to be adopted. The different scanning strategies are used to enhance the isotropic properties in different planes. The strength of the part is comparatively less to the z-direction than the strength in the x-y plane. Considering all these aspects, one can easily use this technology for different applications.

9.4.5 Powder bed fusion

Powder bed fusion is the process that works on the same principle where parts are formed by adding material, shown in Figure 9.7. This technology includes processes such as selective laser melting (SLM), selective laser sintering (SLS), electron beam melting (EBM), and direct metal laser sintering (DMLS) [37]. This process is very famous for using "no" or a "minimum" amount of support structures to build the part. The parts are supported by the surrounding powder, so no additional supports are required during the process. Every part does not require support as it can always be tried to minimize by orientation while printing and with some design modifications as it contributes to waste and can affect the quality of the surface of the printed part. The material is used in powder form on an elevated platform called a "powder bed." The virtual 3D model is sliced into horizontal layers, and the layers of the parts are built-in in 2D form. A thin layer of material powder is delivered on a build platform using a roller or scraper. Energy systems like laser and electron beams are used to sinter or melt the delivered layer of material. Using the energy source, every layer material is selectively fused, and step by step the platform is lowered to complete the building part.

The fusion mechanism understanding is not more superficial as it can be "sintering" and "melting". There are four mechanisms to understand the fusion process. These include complete melting, liquid phase sintering, chemically induced binding, and solid-state sintering. The processing of semi-crystalline polymers and metal alloys follows a complete melting

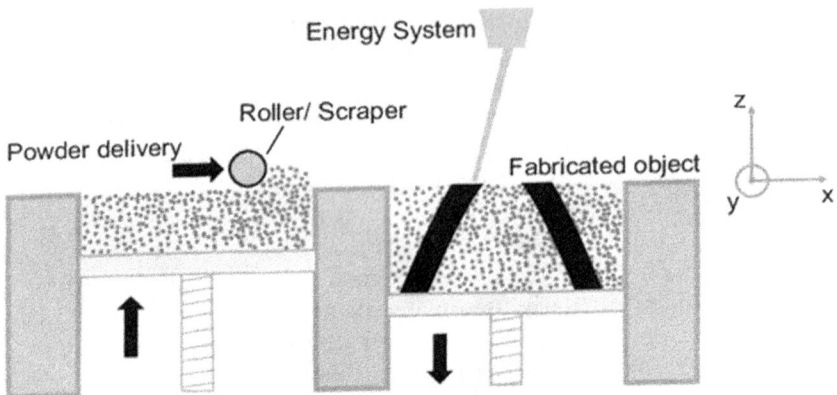

Figure 9.7 Scheme of powder bed fusion process [38]

mechanism. The energy density is enough to melt the material more than layer thickness [39]. The re-melting helps in creating a good bonding of newly formed layers of metals and polymers. Liquid phase sintering is the most common mechanism for major powder bed fusion processes. When applied to the powder, the energy, some portion of the constituents become liquid, and others remain solid; this is also called partial melting. The solid particles are bonded together as some other material is used to act as glue. Chemically induced sintering uses thermally activated chemical reactive material to form powder together. The powder binding can also be done by introducing gasses to the material. Solid-state sintering specifies the fusion of powder particles without melting. The minimization of total free energy, E_s of powder, the particle is the driving force for the solid-state sintering.

$$E_s = \gamma_s \times S_A \tag{9.4}$$

where γ_s is the surface energy per unit area for a particular material, and S_A is the total particle surface area. The particles are closed packed before sintering. They will agglomerate at half of the total melting temperature, minimizing free energy by decreasing surface area, after which porosity decreases as sintering increases [40]. All the material that can be melted and resolidified can be used in this process. The material includes thermoplastic materials, metals, ceramics, and composites.

9.4.6 Sheet lamination

Sheet lamination is the technology used for AM in which thin sheets of materials are bonded together to form a solid part that is modified into a required 3D object [41]. The sheet lamination process is shown in Figure 9.8. The thin sheets are bonded together layer-by-layer using some external energy source. The first commercialized AM technique in this technology was laminated object manufacturing (LOM). Paper material is used with some polymeric adhesives as glue for bonding the multi-layers of paper sheets together. Paper sheets are accurately cut using a laser or mechanical cutter to build the 3D object. Currently, this process is available with different bonding mechanisms like clamping, ultrasonic welding, and thermal bonding. Ultrasonic welding works on solid-state bonding for different applications. The "Bond-then-form," and "Form-then-bond" are two different approaches used as per the limitations of the material and type of part shape required.

The "Bond-then-form" process includes placing the sheet, bonding with the substrate, and cutting according to requirement. The "Form-then-bond" process sheets are already cut to the required shape and bonded with the substrate. This is a more popular approach for various materials and applications as internal features can also be facilitated using this

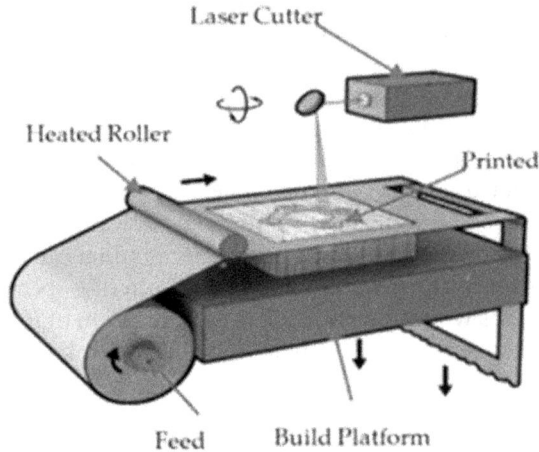

Figure 9.8 Scheme of sheet lamination process [42]

process. LOM is a quick and inexpensive manufacturing technique for large parts. Support structures are not needed to build the part. Using delicate wood-paper-like parts has insufficient strength and may absorb moisture, and achieving a good surface finish is very difficult [43]. This technology is versatile in nature for manufacturing components of a wide range of materials like synthetic material, paper, plastic, fabric material, composites, and metals. As being a cost-effective and automated process, it has particular challenges, such as poor accuracy in z-axis direction and generation of internal features and cavities in the final part. The complex parts that can be achieved using this process are very few. The adjacent layer bonding does not have good binding properties as the internal ones, and that will make binders weaken the interface.

9.4.7 Direct energy deposition

Directed energy deposition is a technology of AM processes that deposits material simultaneously with the heat input. There are different machines available commercially, such as laser engineered net shaping (LENS), laser free form fabrication (LFF), laser cladding, directed light fabrication (DLF), and direct metal deposition (DMD). The heat input can be different such as laser, plasma, and electron beam. Heat energy is used to melt the material to create objects. The feeding of material can be done in the form of powder and wire, shown in Figure 9.9 and Figure 9.10, respectively. The nozzle head moves along the object to deposit the material. Co-axial and off-axis feeding methodologies are used for powder and wire material. In a coaxial nozzle, the heat source is supplied from the center of the nozzle. Material is supplied through channels for interaction with the heat source just before

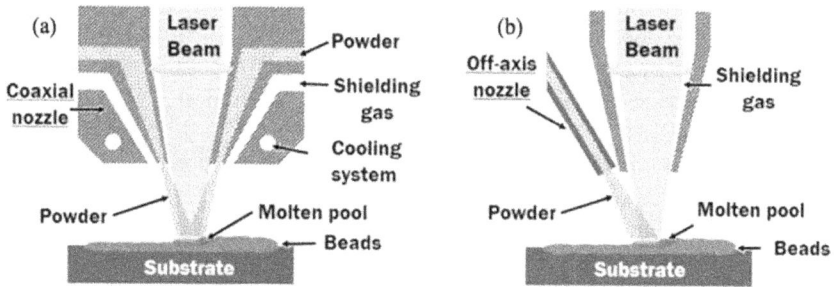

Figure 9.9 Scheme of powder (a) co-axial nozzle (b) off-axis nozzle [47]

the substrate surface. The material is supplied for the interaction angle different from the heat source direction in the off-axis nozzle.

The shielding gas is used to protect the melt pool from the external environment and help in maintaining the cooling rate of the bead microstructure formation. The small melt pool and rapid traverse speed introduce high cooling rates and higher thermal gradients. The high cooling rates produce specific solidification grain structures depending on the material deposited. This technology requires higher energy density and large-size powders that can help in higher deposition rates and less build time. Support structures are not used in this process as multiple axis tunable build platforms are used to achieve different difficult to build features. Based on machine capability, the nozzle can move in multiple directions to deposit the material at any specified angle. The machines like robots are commonly used for this process to improve their ability to build parts in multiple directions [44]. The optimum process parameters are conditional on specific material and its application. These machines are flexible enough to modify and optimize the process parameters. The process parameters are powder feed rate, scan spacing, beam power, spot size, and beam traverse speed. Feed rate and beam power are related to each other, as reducing the feed rate gives a

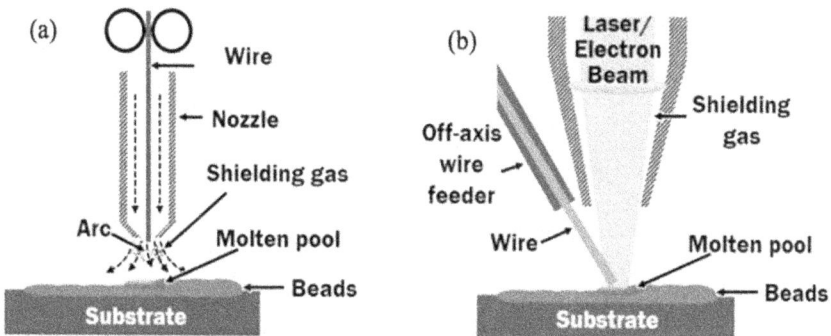

Figure 9.10 Scheme of wire (a) co-axial nozzle (b) off-axis nozzle [47]

similar result to increasing the beam power. The melt pool formation and cooling rate are dependent on the input beam energy [45]. Scan patterns are important as they can be changed after each layer deposition. This changing the orientation minimizes the residual stress formation. Layer height is also an important parameter for selecting heat input as it is necessary to form a melt pool at a depth more than layer thickness to produce the fully dense part. Melt pool size, shape, and temperature are important characteristics of building a sound quality part [46]. Post-processing is needed for the part built by this process, which results in the desired finish of the part. Direct energy deposition systems can be used for printing as well as repairing applications. This method can produce parts of multiple materials at a time. Materials like stainless steel, Inconel 625, nickel alloys, Ti alloys, etc., can be processed to build parts to get the required properties and for various applications.

9.5 DESIGN FOR AM

AM gives freedom to manufacture complicated parts like free-form and intricate features directly from the CAD model. The design for AM (DfAM) parameters is shown in Figure 9.11. There are some design considerations such as lightweight, less residual stress, and the need for support structures, and part orientation is considered for efficient design and ease to manufacture the part.

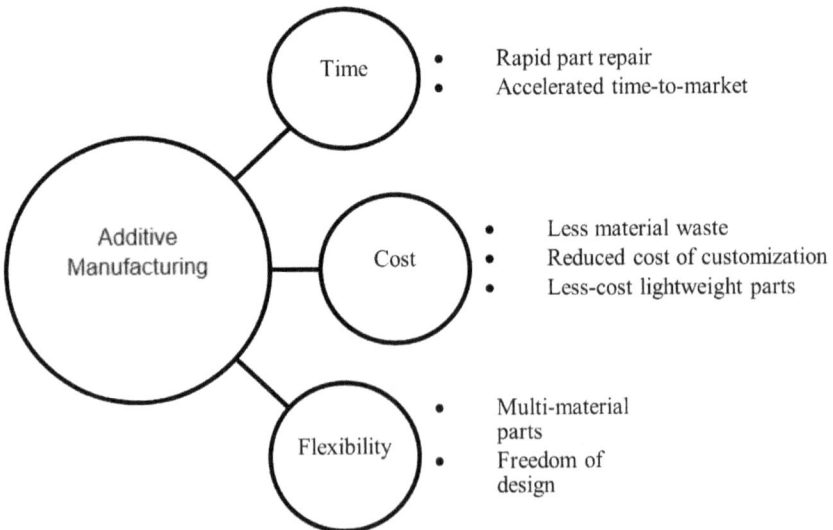

Figure 9.11 DfAM parameters

9.5.1 Residual stress

Residual stress is naturally generated by the heating and cooling of material when applied to the heat source [48]. After every layer, the heat source is applied to melt the material on the surface and fuse the bottom layer. The heat flows down from the melt pool, and molten metal gets cooled and solidified in a very short period. During solidification and cooling of the material, the top layer gets contracted and sets up the shear forces between the top and layer below. This effect can be more critical in parts with large cross-sections as there are longer tracks and more distance on which the shear force can act.

To minimize the residual stress a varying scanning strategy is implemented as per the part geometry [49]. The filling of the material to the center of the part by traveling the heat source is called "hatching." Investigations are conducted to study the effects of scanning strategies on the geometrical and microstructural properties [50]. Different scanning strategies are investigated, including full hatch, full-contour-in-out, and out-in. Hatching patterns help improve build rate and homogenous distribution of residual stress for small and large parts.

9.5.2 Support structures

Support structures are always hard to eliminate even though they will add extra build time and post-processing. Supports have functions to "anchor" the material that is not bonded with the previous layer. This will happen in overhand structures whose angle is less than 45^0 concerning the base platform. Residual stress can be generated in the build if sharp edges are not avoided [51]. Avoiding peeling off the deposited material from the base platform support structure plays an important role. Supports can transfer the heat as unmelted material acts as an insulator. It will help avoid burning, over-melt, and distortion of the deposited material. Supports can be distinguished as primary and secondary supports [52]. Primary supports are generated in the CAD environment along with the main component. These supports can be removed after the completion of the build process. Secondary supports are built with the build setup software as a parameter. Secondary support types include simple supports, angled supports, and tree supports. Removal of the support structure is challenging and may require manual machining. If part geometry is small and weaker, the support can risk damaging the part. Self-supporting features can be introduced to avoid this situation, round holes can be changed to diamond or teardrop shapes with some machining allowance. More points need to be considered while designing the supports, like avoiding tall supports, minimizing overhanging features less than 45^0 to the base platform, and removing areas of horizontal down skin.

9.5.3 Part orientation

Part orientation specifies the tilt and rotation of the part around the axes of machine coordinates. The part production in the build direction is not continuous and an orthogonal shape will be produced continuously. The impact of part orientation is significant, and processes require support structure in the build direction. There are some design rules to be followed for the part orientation. As discussed, a horizontal hole shape needs to be modified to the self-support structure; there should be some minimum value of inner radius for horizontal holes. Orientation can also help improve the quality of the part, dimensional accuracy, building time, shape accuracy, accessibility for support structures, and mechanical strength [53]. The quality of the features depends on the part's geometry, which is oriented by rotation along with part coordinate axes. Overhang features can be printed on an unmelted powder whose thermal conductivity is low compared to solid metal. This results in retaining the heat from the melt pool for a longer time, and the surrounding powder is unnecessarily sintered. This will introduce extra material attached to the part bottom surface of an overhanging region that exhibits a rough finish. Build orientation is finalized at the design stage, and multiple orientation choices are iterated to achieve a minimum amount of required support [54]. It is imperative also to select a proper orientation as it can affect the build time and cost. It is not always easy to orient complex geometry parts, so some priority needs to be given to the build time, cost, support structure, and surface quality.

9.6 FUTURE SCOPE OF AM

Additive manufacturing facilitates the digital design of objects from which physical objects are manufactured. AM can display potential growth in industrial applications for which personal skill power is increasing. The applications of the AM technologies are still in the primary stage of development, and leading industries in the different domains are positively working on adapting this technology. Recent years show different opportunities in AM, referred to as the "big thing." The computers, Internet, and data science introduce the digital journey of AM's technologies. Full-scale production using AM technologies from rapid prototyping is the key opportunity for scaling the development process. AM technologies can save time and money for emergency and life-saving products from the medical field. The expanding areas of AM, such as defense, aerospace, automotive, biomedical, etc., are looking for customized end products. The emerging needs for AM technologies are the demand for novel products for military applications like electronic devices and more functional features in a single part.

These products might be integrated with virtual and physical electronic systems for different analytical and advanced diagnostics in the war field. The wide range of wearable and defense devices use low-cost and multi-functional sensitive sensors [55]. Rapid prototyping and the benefits of AM technology for sensor manufacturing can solve many problems of sensor fabrication. Despite this, the challenges like improving adaptability, performance, and diversity in 3D printed sensors need to be addressed [56]. The ability to build highly complex and lightweight parts with the minimum amount of waste is the reason for AM to make AM technologies a revolution in the aerospace industry. GE Aviation successfully produced the engine fuel nozzle of cobalt chrome material by laser AM technology [57]. NASA engineers produced intricate metal rocket injector parts using AM technology. The existing manufacturing sector of the automobile industry is potentially transforming into value-added and intelligent infrastructure in the era of industrial IoT-based platforms [58]. A higher level of automation and controls can be performed using deep learning to reduce the loads placed on the operator.

AM technologies are addressing critical problems in the medical industry. By eliminating limitations in the current process, process chain development can be implemented involving IoT technologies. The field of reverse engineering, biomaterials, and medicine is involved in solving the problems in the manufacturing of bones and implants [59]. There are different AM application classifications in medicine, including medical models, surgical guides, surgical implants, bio-manufacturing, and external aids. AM plays an important role in reducing the cycle time and rapid production of all products mentioned in the above category. AM technologies can improve the strength of implants, reduce the operation time, provide lightweight parts, fast and low-cost implant production, excellent surface quality, and reconstruction of body parts. The advances in cell printing for artificial blood vessels for bypass surgery and other therapy use AM technology. Exciting applications in the area of surgical planning, medical education and training, and customized implants use AM technologies to achieve potential benefits. Apart from this, in all key fields of transforming industry applications and material challenges, some quality prediction tools are being introduced to predict the design and build using some modeling techniques. Using traditional measurement systems, it is very difficult to measure internal cavities, channels, undercuts, and porosity. For internal geometries, x-ray computer tomography can be a solution. The key problems of inspecting AM parts are related to uncertainty and system calibration. It is difficult for AM to take over the entire manufacturing industry, but all can predict that AM has a great market and growth in the near future. A summary of AM technologies is provided in Table 9.1.

Table 9.1 Summary of AM technologies

TECHNOLOGY	Vat Polymerization	Material Jetting	Binder Jetting	Material Extrusion	Powder Bed Fusion	Sheet Lamination	Direct Energy Deposition
PROCESS	Stereolithography (SLA)	Polyjet Printing	Indirect Inkjet Printing	Fused Deposition Modeling (FDM)	SLS DMLS SLM EBM	Laminated Object Manufacturing (LOM)	Laser Engineered Net Shape (LENS)
MATERIAL	Photopolymer Ceramics	Wax, Photopolymer	Polymer-Metal, Ceramic Powder	Thermoplastic Metal Paste	Metal, Ceramics Powder	Paper, Plastic Films, Metal Sheets	Metal Powder and Wire
PROS	Good Part Resolution High printing Speed	Multi Material Printing High Surface Finish	Material Available Color Parts	Multi Material Printing Rapid Building	High Accuracy and Details High strength Parts	High Surface Finish	Repair of Damaged Parts Functionally Graded Materials
CONS	High Material Cost	Less Strength of Parts	High Porosity	Poor Surface Finish	Powder Handling	Decubing Issues	Expensive Bed Resolution

9.7 CONCLUSION

AM is the process that can help individuals and industries build physical parts using CAD models by sequentially adding material layer by layer. The unique abilities of AM technologies open new opportunities for researchers to explore new challenges in the design of products for improving part performance. Many considerations can illustrate the design solution of the customized products. Additive manufacturing can print the assembly of a product in a single piece, saving money and time. The progress in software development by many technology companies is meeting the requirements of AM technologies to solve different problems. All the AM technology has its pros and cons that can help select the appropriate process to meet the requirements of properties and functional capabilities of the product. Industry 4.0 includes digital technologies to make AM more flexible, responsive, and agile to a wide range of customers.

REFERENCES

1. A. K. Sharma, R. Bhandari, C. Pinca-Bretotean, C. Sharma, S. K. Dhakad, and A. Mathur, "A study of trends and industrial prospects of Industry 4.0," *Mater. Today Proc.*, vol. 47, no. xxxx, pp. 2364–2369, 2021, doi: 10.1016/j.matpr.2021.04.321.
2. E. Puskás and G. Bohács, "Physical internet – A novel application area for industry 4.0," *Int. J. Eng. Manag. Sci.*, vol. 4, no. 1, pp. 152–161, 2019, doi: 10.21791/ijems.2019.1.19.
3. R. Ashima, A. Haleem, S. Bahl, M. Javaid, S. K. Mahla, and S. Singh, "Automation and manufacturing of smart materials in additive manufacturing technologies using internet of things towards the adoption of industry 4.0," *Mater. Today Proc.*, vol. 45, pp. 5081–5088, 2021, doi: 10.1016/j.matpr.2021.01.583.
4. K. Salonitis, "Design for additive manufacturing based on the axiomatic design method," *Int. J. Adv. Manuf. Technol.*, vol. 87, no. 1–4, pp. 989–996, 2016, doi: 10.1007/s00170-016-8540-5.
5. A. Ceruti, P. Marzocca, A. Liverani, and C. Bil, "Maintenance in aeronautics in an industry 4.0 context: The role of augmented reality and additive manufacturing," *J. Comput. Des. Eng.*, vol. 6, no. 4, pp. 516–526, 2019, doi: 10.1016/j.jcde.2019.02.001.
6. A. Equbal, S. Akhter, A. K. Sood, and I. Equbal, "The usefulness of additive manufacturing (AM) in COVID-19," *Ann. 3D Print. Med.*, vol. 2, p. 100013, 2021, doi: 10.1016/j.stlm.2021.100013.
7. R. Minetto, N. Volpato, J. Stolfi, R. M. M. H. Gregori, and M. V. G. da Silva, "An optimal algorithm for 3D triangle mesh slicing," *CAD Comput. Aided Des.*, vol. 92, pp. 1–10, 2017, doi: 10.1016/j.cad.2017.07.001.
8. J. Huang, Q. Qin, and J. Wang, "A review of stereolithography: Processes and systems," *Processes*, vol. 8, no. 9, 2020, doi: 10.3390/PR8091138.

9. H. Gojzewski *et al.*, "Layer-by-layer printing of photopolymers in 3D: How weak is the interface?," *ACS Appl. Mater. Interfaces*, vol. 12, no. 7, pp. 8908–8914, 2020, doi: 10.1021/acsami.9b22272.

10. Y. Yang *et al.*, "Three dimensional printing of high dielectric capacitor using projection based stereolithography method," *Nano Energy*, vol. 22, pp. 414–421, 2016, doi: 10.1016/j.nanoen.2016.02.045.

11. G. Strano, L. Hao, R. M. Everson, and K. E. Evans, "A new approach to the design and optimisation of support structures in additive manufacturing," *Int. J. Adv. Manuf. Technol.*, vol. 66, no. 9–12, pp. 1247–1254, Jun. 2013, doi: 10.1007/S00170-012-4403-X.

12. Z. Chen *et al.*, "3D printing of piezoelectric element for energy focusing and ultrasonic sensing," *Nano Energy*, vol. 27, pp. 78–86, 2016, doi: 10.1016/j.nanoen.2016.06.048.

13. D. Lin *et al.*, "3D stereolithography printing of graphene oxide reinforced complex architectures," *Nanotechnology*, vol. 26, no. 43, p. 434003, 2015, doi: 10.1088/0957-4484/26/43/434003.

14. A. Andreu *et al.*, "4D printing materials for vat photopolymerization," *Addit. Manuf.*, vol. 44, no. May, p. 102024, 2021, doi: 10.1016/j.addma.2021.102024.

15. S. Tibbits, "4D printing: Multi-material shape change," *Archit. Des.*, vol. 84, no. 1, pp. 116–121, 2014, doi: 10.1002/ad.1710.

16. S. S. Ray, H. Dommati, J. C. Wang, and S. S. Chen, "Solvent based Slurry Stereolithography 3D printed hydrophilic ceramic membrane for ultrafiltration application," *Ceram. Int.*, vol. 46, no. 8, pp. 12480–12488, 2020, doi: 10.1016/j.ceramint.2020.02.010.

17. M. Casafont, J. M. Pons, J. Bonada, M. M. Pastor, F. Marimon, and F. Roure, "Experimental study of the compression behavior of mask image projection based on stereolithography manufactured parts," *Procedia Manuf.*, vol. 41, pp. 460–467, 2019, doi: 10.1016/j.promfg.2019.09.033.

18. H. Liu, N. Yang, Y. Sun, L. Yang, and N. Li, "Effect of the build orientation on the mechanical behaviour of polymers by stereolithography," *IOP Conf. Ser. Mater. Sci. Eng.*, vol. 612, no. 3, 2019, doi: 10.1088/1757-899X/612/3/032166.

19. D. L. Naik and R. Kiran, "On anisotropy, strain rate and size effects in vat photopolymerization based specimens," *Addit. Manuf.*, vol. 23, no. June, pp. 181–196, 2018, doi: 10.1016/j.addma.2018.08.021.

20. T. Hada, M. Kanazawa, M. Iwaki, T. Arakida, and S. Minakuchi, "Effect of printing direction on stress distortion of three-dimensional printed dentures using stereolithography technology," *J. Mech. Behav. Biomed. Mater.*, vol. 110, no. May, p. 103949, 2020, doi: 10.1016/j.jmbbm.2020.103949.

21. E. Willems *et al.*, "Additive manufacturing of zirconia ceramics by material jetting," *J. Eur. Ceram. Soc.*, vol. 41, no. 10, pp. 5292–5306, 2021, doi: 10.1016/j.jeurceramsoc.2021.04.018.

22. E. Salcedo, D. Baek, A. Berndt, and J. E. Ryu, "Simulation and validation of three dimension functionally graded materials by material jetting," *Addit. Manuf.*, vol. 22, no. May 2017, pp. 351–359, 2018, doi: 10.1016/j.addma.2018.05.027.

23. R. MacCurdy, J. Lipton, S. Li, and D. Rus, "Printable programmable viscoelastic materials for robots," *2016 IEEE/RSJ International Conference on Intelligent Robots and Systems (IROS)*, vol. 2016, pp. 2628–2635, 2016, doi: 10.1109/IROS.2016.7759409.

24. E. L. Doubrovski, E. Y. Tsai, D. Dikovsky, J. M. P. Geraedts, H. Herr, and N. Oxman, "Voxel-based fabrication through material property mapping: A design method for bitmap printing," *CAD Comput. Aided Des.*, vol. 60, pp. 3–13, 2015, doi: 10.1016/j.cad.2014.05.010.

25. P. Gay, D. Blanco, F. Pelayo, A. Noriega, and P. Fernández, "Analysis of factors influencing the mechanical properties of flat PolyJet manufactured parts," *Procedia Eng.*, vol. 132, pp. 70–77, 2015, doi: 10.1016/j.proeng.2015.12.481.

26. A. Ion *et al.*, "Metamaterial mechanisms," *UIST '16: Proceedings of the 29th Annual Symposium on User Interface Software and Technology*, pp. 529–539, 2016, doi: 10.1145/2984511.2984540.

27. A. Pugalendhi, R. Ranganathan, and S. Ganesan, "Impact of process parameters on mechanical behaviour in multi-material jetting," *Mater. Today Proc.*, vol. 46, pp. 9139–9144, 2019, doi: 10.1016/j.matpr.2019.12.106.

28. J. Zhang, B. J. Allardyce, R. Rajkhowa, X. Wang, and X. Liu, "3D printing of silk powder by binder jetting technique," *Addit. Manuf.*, vol. 38, p. 101820, Feb. 2021, doi: 10.1016/J.ADDMA.2020.101820.

29. T. Tancogne-Dejean, C. C. Roth, and D. Mohr, "Rate-dependent strength and ductility of binder jetting 3D-printed stainless steel 316L: Experiments and modeling," *Int. J. Mech. Sci.*, vol. 207, p. 106647, Oct. 2021, doi: 10.1016/J.IJMECSCI.2021.106647.

30. M. Sireesha, J. Lee, A. S. Kranthi Kiran, V. J. Babu, B. B. T. Kee, and S. Ramakrishna, "A review on additive manufacturing and its way into the oil and gas industry," *RSC Adv.*, vol. 8, no. 40, pp. 22460–22468, 2018, doi: 10.1039/c8ra03194k.

31. S. Manotham and P. Tesavibul, "Effect of particle size on mechanical properties of alumina ceramic processed by photosensitive binder jetting with powder spattering technique," *J. Eur. Ceram. Soc.*, vol. 42, no. 4, pp. 1608–1617, 2022, doi: 10.1016/j.jeurceramsoc.2021.11.062.

32. N. Lecis *et al.*, "Effects of process parameters, debinding and sintering on the microstructure of 316L stainless steel produced by binder jetting," *Mater. Sci. Eng. A*, vol. 828, no. July, p. 142108, 2021, doi: 10.1016/j.msea.2021.142108.

33. H. Deng, Y. Huang, S. Wu, and Y. Yang, "Binder jetting additive manufacturing: Three-dimensional simulation of micro-meter droplet impact and penetration into powder bed," *J. Manuf. Process.*, vol. 74, no. July 2021, pp. 365–373, 2022, doi: 10.1016/j.jmapro.2021.12.019.

34. R. Patel, C. Desai, S. Kushwah, and M. H. Mangrola, "A review article on FDM process parameters in 3D printing for composite materials," *Mater. Today Proc.*, Feb. 2022, doi: 10.1016/J.MATPR.2022.02.385.

35. M. E. Lamm *et al.*, "Material extrusion additive manufacturing of wood and lignocellulosic filled composites," *Polymers (Basel).*, vol. 12, no. 9, 2020, doi: 10.3390/POLYM12092115.

36. M. Qamar Tanveer, G. Mishra, S. Mishra, and R. Sharma, "Effect of infill pattern and infill density on mechanical behaviour of FDM 3D printed Parts- a current review," *Mater. Today Proc.*, Mar. 2022, doi: 10.1016/J.MATPR.2022.02.310.

37. P. Avrampos and G. C. Vosniakos, "A review of powder deposition in additive manufacturing by powder bed fusion," *J. Manuf. Process.*, vol. 74, pp. 332–352, Feb. 2022, doi: 10.1016/J.JMAPRO.2021.12.021.

38. Wiberg, A., "Towards Design Automation for Additive Manufacturing A Multidisciplinary Optimization approach," Linköping Studies in Science., p. 69, 2019.
39. G. V. de Leon Nope, L. I. Perez-Andrade, J. Corona-Castuera, D. G. Espinosa-Arbelaez, J. Muñoz-Saldaña, and J. M. Alvarado-Orozco, "Study of volumetric energy density limitations on the IN718 mesostructure and microstructure in laser powder bed fusion process," *J. Manuf. Process.*, vol. 64, pp. 1261–1272, Apr. 2021, doi: 10.1016/J.JMAPRO.2021.02.043.
40. F. Chu *et al.*, "Influence of satellite and agglomeration of powder on the processability of AlSi10Mg powder in laser powder bed fusion," *J. Mater. Res. Technol.*, vol. 11, pp. 2059–2073, Mar. 2021, doi: 10.1016/J.JMRT.2021.02.015.
41. Y. Y. Chiu, Y. S. Liao, and C. C. Hou, "Automatic fabrication for bridged laminated object manufacturing (LOM) process," *J. Mater. Process. Technol.*, vol. 140, no. 1–3, pp. 179–184, Sep. 2003, doi: 10.1016/S0924-0136(03)00710-6.
42. A. Vafadar, F. Guzzomi, A. Rassau, and K. Hayward, "Advances in metal additive manufacturing: A review of common processes, industrial applications, and current challenges," *Appl. Sci.*, vol. 11, no. 3, pp. 1–33, 2021, doi: 10.3390/app11031213.
43. D. Klosterman, R. Chartoff, G. Graves, N. Osborne, and B. Priore, "Interfacial characteristics of composites fabricated by laminated object manufacturing," *Compos. Part A Appl. Sci. Manuf.*, vol. 29, no. 9–10, pp. 1165–1174, Jan. 1998, doi: 10.1016/S1359-835X(98)00088-8.
44. B. Freire, M. Babcinschi, L. Ferreira, B. Señaris, F. Vidal, and P. Neto, "Direct energy deposition: A complete workflow for the additive manufacturing of complex shape parts," *Procedia Manuf.*, vol. 51, pp. 671–677, Jan. 2020, doi: 10.1016/J.PROMFG.2020.10.094.
45. C. Vundru, R. Singh, W. Yan, and S. Karagadde, "Effect of spreading of the melt pool on the deposition characteristics in laser directed energy deposition," *Procedia Manuf.*, vol. 53, pp. 407–416, Jan. 2021, doi: 10.1016/J.PROMFG.2021.06.043.
46. Z. Wang *et al.*, "Influence of powder characteristics on microstructure and mechanical properties of Inconel 718 superalloy manufactured by direct energy deposition," *Appl. Surf. Sci.*, vol. 583, p. 152545, May 2022, doi: 10.1016/J.APSUSC.2022.152545.
47. D. G. Ahn, *Directed Energy Deposition (DED) Process: State of the Art*, vol. 8, no. 2. Korean Society for Precision Engineering, 2021.
48. S. Waqar, K. Guo, and J. Sun, "Evolution of residual stress behavior in selective laser melting (SLM) of 316L stainless steel through preheating and in-situ re-scanning techniques," *Opt. Laser Technol.*, vol. 149, p. 107806, May 2022, doi: 10.1016/J.OPTLASTEC.2021.107806.
49. N. Nadammal *et al.*, "Critical role of scan strategies on the development of microstructure, texture, and residual stresses during laser powder bed fusion additive manufacturing," *Addit. Manuf.*, vol. 38, p. 101792, 2021, doi: 10.1016/j.addma.2020.101792.
50. C. A. Biffi *et al.*, "Effects of the scanning strategy on the microstructure and mechanical properties of a TiAl6V4 alloy produced by electron beam additive manufacturing," *Int. J. Adv. Manuf. Technol.*, vol. 107, no. 11–12, pp. 4913–4924, 2020, doi: 10.1007/s00170-020-05358-y.

51. S. Sunny, R. Mathews, H. Yu, and A. Malik, "Effects of microstructure and inherent stress on residual stress induced during powder bed fusion with roller burnishing," *Int. J. Mech. Sci.*, vol. 219, p. 107092, Apr. 2022, doi: 10.1016/J.IJMECSCI.2022.107092.

52. Z. Wang, Y. Zhang, S. Tan, L. Ding, and A. Bernard, "Support point determination for support structure design in additive manufacturing," *Addit. Manuf.*, vol. 47, p. 102341, Nov. 2021, doi: 10.1016/J.ADDMA.2021.102341.

53. B. Torries, N. Shamsaei, and S. Thompson, "Effect of build orientation on fatigue performance of Ti-6Al-4V parts fabricated via laser-based powder bed fusion," *Solid Freeform Fabrication 2017: Proceedings of the 28th Annual International Solid Freeform Fabrication Symposium – An Additive Manufacturing Conference*, pp. 115–121, 2020.

54. E. Tan Zhi'En, J. H. L. Pang, and J. Kaminski, "Directed energy deposition build process control effects on microstructure and tensile failure behaviour," *J. Mater. Process. Technol.*, vol. 294, p. 117139, Aug. 2021, doi: 10.1016/J.JMATPROTEC.2021.117139.

55. N. Afsarimanesh, A. Nag, S. Sarkar, G. S. Sabet, T. Han, and S. C. Mukhopadhyay, "A review on fabrication, characterization and implementation of wearable strain sensors," *Sensors Actuators, A Phys.*, vol. 315, p. 112355, 2020, doi: 10.1016/j.sna.2020.112355.

56. U. Shamsi, "Monitoring applications," *GIS Appl. Water, Wastewater, Stormwater Syst.*, 2005, doi: 10.1201/9781420039252.ch10.

57. D. I. Wimpenny, P. M. Pandey, and L. Jyothish Kumar, "Advances in 3D printing & additive manufacturing technologies," *Adv. 3D Print. Addit. Manuf. Technol.*, pp. 1–186, 2016, doi: 10.1007/978-981-10-0812-2.

58. L. Haghnegahdar, S. S. Joshi, and N. B. Dahotre, "From IoT-based cloud manufacturing approach to intelligent additive manufacturing: Industrial internet of things—An overview," *Int. J. Adv. Manuf. Technol.*, pp. 1461–1478, 2022, doi: 10.1007/s00170-021-08436-x.

59. M. Javaid and A. Haleem, "Additive manufacturing applications in medical cases: A literature based review," *Alexandria J. Med.*, vol. 54, no. 4, pp. 411–422, 2018, doi: 10.1016/j.ajme.2017.09.003.

Chapter 10

Recycling of metal/polymer waste as a feedstock material via additive manufacturing technology

A review

Dipak A. Patil, Malkeet Singh, Shilpi Chaudhary, and Harpreet Singh

10.1 INTRODUCTION

In recent years, additive manufacturing (AM) technology has evolved rapidly and gradually shifted the traditional application method's focus. The development of AM techniques is becoming popular in the manufacturing industry. AM is defined by ASTM F2792-12a (Standard Terminology for Additive Manufacturing Technologies) as the "process of joining materials to make objects from 3D model data, usually layer upon layer, as opposed to subtractive manufacturing methodologies, such as traditional machining" [1].

Additive manufacturing was initially used for producing 3D prototypes and models of concept by researchers and engineers. Hence it is also popularly known as rapid prototyping. Later, with the AM techniques' advancement, its use is spread in broad sectors such as automobile, aerospace, bio-medical sector, tooling industry, toy industry, architecture and art field, consumer goods, and decorative products [2]. It is also known by names like solid free-forming, rapid manufacturing, rapid tooling, layer manufacturing, and direct digital manufacturing [3].

Utilizing the AM techniques in different sectors made it possible to develop complex, innovative components and parts. But the materials available for 3D printing are limited and more research needs to be done to find out more novel materials. The optimum and cost-effective use of the AM techniques called for the development of new multi-functional materials [4]. Recycling the material, for example, polymer waste and metal scrap, and using it as a feedstock material for the 3D printing technique is a potential option. The polymer is the most widely used material and applicable for domestic purposes and various manufacturing sectors. The polymer became very popular in the manufacturing industry because of its desirable properties like its light weight, ease to manufacture, and durability [5, 6]. The use of plastic has risen over 500 percent in the last

DOI: 10.1201/9781003246466-10

three decades and continuously increases, leading to the waste disposal problem [7]. Plastic does not degrade quickly; it takes around 500 to 1,000 years to decompose depending on the type of plastic used. Packaging items, for example, trays, cans, bottles, containers, produce plastic waste in huge amounts. This waste plastic acquires a large land area that can also be used for other valuable purposes [8]. It has a harmful impact on the environment. Many incidents happen as animals die because of eating plastic and polymers. Aqua life is poorly affected because of the polymer/plastic waste thrown in the sea and river. Researchers made attempts for recycling the polymer/plastic because of the day by day increase of polymer/plastic waste. Plastic is a valuable resource, and its optimum use can be done by recycling [9]. Recycling is collecting the polymer/plastic for its recovery, reprocessing it, and using it differently. The number of factors such as the type of plastic, the composition of polymer and additives in the plastic, and the number of times it has previously been recycled, needs to be considered while recycling the plastic. Hence plastic recycling is a tedious process [10].

Plastic is recycled mainly in two ways: mechanical recycling and chemical recycling. Mechanical recycling in which plastic is collected, separated, washed to remove contaminants, ground to the small particles, melted, and solidified. Mechanical recycling is popularly used in the industry for the recycling of polymers. Plastic such as nylon, polyethylene terephthalate, and polyurethane can be recycled by chemical recycling. In chemical recycling, the polymer is broken down and depolymerized into monomers. Depolymerization of the polymer can be done by various processes like hydrolysis, glycolysis, and methanolysis [11]. The work done on plastic recycling and prior methods is documented in the present chapter.

Metal is the material that is used more commonly in AM. One of the main challenges in the metal AM techniques is the cost of 3D printed components, and material cost is a significant factor in the process's total cost [12]. Hence finding the alternative for reducing the initial metal powder cost and utilizing it entirely in the AM techniques becomes the important parameter for sustainable AM. Recycling the industrial metal scrap and machining chips and using them as feedstock material in AM can be a potential option for cost-effective metal AM. Work needs to be done to develop efficient and cost-effective strategies for recycling different metallic waste.

The development of efficient and cost-effective strategies for the recycling of polymer and industrial metal waste is a key area for research. The present chapter gives an overview of the work done on recycling the polymer packaging waste and metal scrap by using them as feedstock material for the AM technique.

10.2 STEPS INVOLVED, ADVANTAGES, AND LIMITATIONS OF AM

Making 3D-shaped parts using AM technique can be divided into three main steps: modeling, printing, and post-processing. The detailed steps involved in 3D printing are as follows. In the first step, the 3D CAD model is prepared using software like SolidWorks, AutoCAD, FreeCAD, OpenSCAD, Inventor, or the object can be scanned by a 3D scanner like a CT scan or MRI.

The STL is created then processed through slicer packages like Slic3r, KISSlicer, Repetier, and Cura. Slicer package divides the model into a number of 2D layers, and according to each layer shape, it prepares the G code for the printer. In the second step, the prepared G code is executed by the printer, and a 3D model is created by the deposition of layer by layer under the computer control. In the last step, post-processing can be done to get more accurate tolerances. Usually, the part printed has a nice finish and tolerance, suitable for most standard applications [5].

AM is more advantageous than conventional manufacturing processes. Any part can be directly printed, and no tooling and fixture are required. Hence it eliminates the lead time of the process and so on, which eliminates the tooling cost, setup cost, assembly cost, and post-processing cost of the process. Complex shapes and higher geometrical tolerance products, which are difficult to make using conventional techniques, can easily be made with AM techniques. Also, any changes in geometry can be easily incorporated by making changes in the 3D CAD model. It gives them the freedom to make customized products.

Though AM offers lots of advantages, it has some limitations as well. The time needed and cost for printing are the prime concerns. The time required for the printing can be from hours to days, depending on the product's size, geometry, and complexity [13].

The higher cost of the process is another factor, and feedstock cost contributes greatly to the overall cost of the process. Also, a limited number of materials are available for printing. Other limitations are that the size of the part produced is constrained by the build area of the printer [14], the void formation, anisotropy in the mechanical properties, and microstructure [15].

10.3 ADDITIVE MANUFACTURING PROCESSES

10.3.1 Selective laser sintering (SLS)

In this process of AM, a laser is utilized as a power source to sinter the feedstock powdered material in the selective laser sintering (SLS) process. The SLS method commonly uses metal powder as a feedstock material. The roller spreads a consistent layer of height equal to one layer across the build platform during the printing process. The laser uses a 3D model to point at a specific area and selectively sinters the powder. Once the first layer has been printed, the powder is distributed over the first sintered layer with a roller, and the procedure is repeated until the entire 3D model has been created. After printing, the printed model is removed and cleaned. SLS is an emerging technology and is limited to low-volume production [13].

10.3.2 Fused deposition melting (FDM)

FDM is one of the most extensively utilized 3D printing processes. Plastic is typically used as a feedstock in this process. The build platform, extruder nozzles, feedstock spool, and liquefiers (heater) are the main part of this method's setup. The plastic filament is fed from the feedstock spool to the heated extruder during printing. The heated extruder melts the plastic filament, which emerges from the extrusion nozzle's tip. The extruder can move in two directions (X-direction and Y-direction) while the build platform moves up and down (Z-direction). The nozzle deposits melted material on the build platform in the initial layer of printing. In the first layer of the printing, the nozzle deposits the melted material on the build platform. The extruder moves as per the first layer geometry of the part. After depositing the first layer, the build platform is lowered down by height equal to layer height and this process continues to form a 3D modeled part. Stepper motors are generally used to give the motion in all three axes [13, 14].

10.3.3 Stereolithography

Stereolithography uses UV light or a laser beam to cure the photosensitive polymers. In this process, the liquid photopolymer is spread in layers by a sweeper. As per the geometry of each layer, the UV rays polymerize the liquid photosensitive polymer. Various lenses are used in the process to focus the rays, and scanning mirrors are used to point the laser beam at the specific location on the build platform. As the liquid polymer is used, which can be accurately polymerized by laser beam, the 3D printed model produced by this method gives good part tolerance and surface finish [14].

10.3.4 Binder jetting

In the process of binder jetting, powder form materials are used for building the 3D objects, while the binder is used as adhesive material between the two layers of powder material. The powder is spread over the platform by a roller sweeper. The inkjet print head moves over the platform and injects the binder at a specific location as per the layer geometry. Once the first layer is completed the sweeper again spreads the powder over the previous layer and the second layer powder particle gets stuck to the first layer particle because of the adhesive binder in between them. In this way, alternate layers of powder and binder are deposited to form the final 3D object. This process can be used for the powder form material of metals, polymers, and ceramics [13, 14].

10.3.5 Laminated object manufacturing

Laminated object manufacturing is a new emerging technology of additive manufacturing and is still in development. In this technique, the sheets of paper, plastic, or metals are stuck one over another to form the object. A feed roll supplies the sheet and these adhesive coated sheets are cut in the required shape using the laser cutter. This cutting process goes on and each layer gets joined with the previous layer. The final object form may need post-processing machining to get better shape and dimensional accuracy [16].

10.4 COLD SPRAY PROCESS

In the cold spray process, a supersonic compressed gas jet accelerates powder particles to very high velocities at temperatures below their melting point. De-Laval converging-diverging nozzle is used to accelerate the gas to supersonic velocity. Helium and nitrogen are generally used as carrier gas in the cold spray process. Once the gasses achieve the supersonic velocity, the metal powder is introduced into the flow and with the interaction with gas flow, powder particles also get accelerated to a higher velocity. These accelerated powder particles are made to strike on the substrate on which they get plastically deformed and form the layer. The successive particles get deposited one over the other to form the objects. Particles should have a higher velocity than critical velocity. Critical velocity is the minimum velocity which a particle needs to attend to get plastically deformed on the impact on the substrate. The cold spray also finds its application in coatings and the maintenance of worn-out parts. As a result, the cost of the initial setup, as well as the powder and gas consumed, are higher. The cost of the printed part is higher with this technology [16].

10.5 COMMON MATERIALS USED IN AM

Initially, only polymer material was available for the 3D printing process. With the development of new AM techniques, a variety of materials like metal, composites, and ceramics can be used to develop parts [16]. This has also made possible the use of various multi-functional materials for AM. Polymer and metal are used more popularly in AM.

10.5.1 Polymers

The polymer is one of the most popular materials used in additive manufacturing as it offers desirable properties, and processing plastic with the help of heat or light is easy and many AM methods like selective laser sintering (SLS), fused deposition modeling (FDM), and stereolithography are available for the 3D printing of the plastic components. FDM is most commonly used for a polymer having low melting points [17]. With the continuous development in the technologies and new novel materials, polymers and composites are finding wide applications in various sectors. Less energy conservation, less waste, in-house production, and ease in customizing product design are important aspects in using polymer materials in the AM techniques. The ability to make complex shape geometry and light-weight parts are important specifications in the aerospace and automobile sectors and they can be easily achieved with the development of polymer parts using AM techniques. The polymer has been popularly used in the toy industry for a long time. With the AM techniques, it is now possible to give sufficient details and complex shapes to the toy [18]. Also, a polymer is gaining importance in applications such as making architectural and historical sculptures, making implants, prostheses in the biomedical sectors, rapid tooling, and consumer goods products.

Plastic (polymer) products are used extensively in all sectors. After the end of useful life, the product is thrown away as waste. Packaging items like trays, cans, bottles, and containers also produce plastic waste in huge amounts. Due to the increasing plastic waste, attempts were made by researchers to recycle plastic, as plastic is a significant resource and its optimum use can be achieved by recycling.

In recent years, researchers made many attempts to recycle plastic and make a feedstock material for 3D printing. Polymers used mainly in the toy industry include polyethylene, polystyrene, polypropylene, acrylonitrile butadiene styrene (ABS), and polyvinyl chloride [18]. Usually, after breaking or when it stops functioning, it becomes a scrap, but the material it is made up of can be reused. Hence recycling such toys is a good option. Nur-A-Tomal et al. [19] used waste plastic toys. ABS single color toys were shredded, and test specimens were printed using a single screw extruder. Tensile and Charpy impact testing was carried out to evaluate mechanical properties such as strength, strain, and modulus of the printed specimen. This test showed comparable results between the printed specimen and pure material in terms of mechanical properties. ABS and polylactic acid (PLA) are the most commonly used in 3D printing [20]. ABS is a thermoplastic polymer very popularly used in the manufacturing industry because of its desirable properties such as high impact resistance, rigidity, light weight, and value. Also, it offers opaque, smooth, and shiny surfaces in 3D printing. In the case of typical plastic used in the FDM, a significant volume of material is lost due to the use of rafts, supports, and failed prints. In such cases, for the optimum use of material, recycling becomes an important aspect. Mazher Iqbal Mohammed et al. [21] used waste 3D prints and raft/support material and virgin pallet material of ABS to prepare the filament. 3D printer filament made from both granulated and pallet material of ABS was used to print two different 3D parts using the FDM printer. A comparative analysis was performed on both samples to examine the recycled ABS material's printability. Degradation of mechanical properties was observed, which resulted in a 13 to 49 percent decrease in ultimate tensile strength. However, the recycled filament's good print quality was observed as consistent with commercial filament, which was confirmed by surface roughness analysis.

PLA is one of the most important polymers. Many common use articles and packaging materials such as containers, trays, and bottles are made from the PLA [22, 23]. It is also a biodegradable and biocompatible polymer and hence is used in the biomedical sector [24]. After the useful life of these components, they can be recycled to reduce environmental impact and resource depletion. PLA is one of the most used polymer materials for FDM filament; hence many researchers have evaluated the potential of using a recycled polymer as a filament material. Fabio Cruz et al. [25] evaluated the mechanical recyclability of PLA, which was used in the open-source 3D printing and the potential to use recycled PLA as filament material. The author observed that the elastic domain strength of material remained constant, but strain at break reduced from 1.88 percent to 1.68 percent. The molecular weight of recycled material decreased strongly after the recycling process. It was observed that molecular weight reduction of about 26.73 percent was achieved after three cycles with respect to virgin polymer, and 46.91 percent was achieved after the five cycles. The rheological experiment showed a drop in viscosity. Zero shear viscosity decreased from 2729.21 Pa. s to 219.85 Pa. s when PLA extruded five times. Melt flow index (MFI) shows an increase of 6.05 times compared to virgin value after the recycling process.

Daniel Tanney et al. [26] investigated the potential of recycling the waste generated in prototyping and change in the mechanical properties of recycled PLA with the number of life cycles it has gone through in material extrusion machines. They also investigated the effect of adding virgin pellets to pellets of recycled material in different ratios on the different properties, like tensile strength and impact strength. In the case of PLA, over ten extrusions cycles, material properties almost remained constant while tensile strength and impact strength of the material was reduced by 5.2 percent and 20.2 percent, respectively. The addition of virgin pellets to recycled pallets increased the tensile strength of filament in the initial recycling stages while the strain at failure and toughness remained the same by the addition of virgin polymer.

Zenkiewicz et al. [27] carried out the multiple extrusion cycles on PLA and evaluated the effect of recycling on the mechanical properties, impact strength, melt flow index (MFI), glass transition and phase transition temperature, and thermal stability. Degradation in mechanical properties was observed with an increase in the number of extrusion cycles. The variation of mechanical properties as a function of a number of extrusion cycles was also discussed in detail by the investigator.

After ten extrusion cycles, a decrease in tensile strength by 5.2 percent, tensile strength at the break by 8.3 percent was observed. The impact strength was decreased by 20.2 percent. After ten extrusion cycles, MFI was increased by 236 percent compared to its original value. The increasing trend of MFI as a function of extrusion number. The number of extrusions did not affect the glass transition of PLA while the cold crystallization temperature decreased with an increase in the number of extrusion cycles. Thus, it is clear that the mechanical properties and strength of the polymer decrease with recycling. In addition to that, warpage/shrinkage in the 3D printed component also increases compared to virgin material used in the 3D printing processes [28]. This distortion can be reduced and the stiffness of the part can be increased by adding reinforcement materials. The addition of cellulose fiber in the polypropylene increases the strength of the recycled polymer and decreases the thermal expansion. It provides thermal dimensional stability [29].

Nicole E. Zander et al. [30] presented work focused on reinforcing recycled polypropylene with the use of cellulose waste materials to produce a green composite feedstock for extrusion-based polymer additive manufacturing. They have used easily available cellulose materials from waste paper, cardboard, and wood flour and added propylene using the solid-state shear pulverization process. Filaments were made using this powder and used in the extrusion additive manufacturing process. Test specimens were printed using recycled and commercial polypropylene (PP) materials. Results from scanning electron microscope (SEM) and thermogravimetric (TGA) analysis showed higher loading of cellulose in recycled PP compared to commercial PP. From the dynamic mechanical analysis, it was observed that storage modulus increased 20 to 30 percent on the addition of cellulose materials. In addition to 10 wt percent of cellulose in virgin polypropylene, elastic modulus increased by 38 percent. Studies were also carried out to improve the mechanical properties and thermal characteristics of the recycled polymer by adding metal particles to it.

Seyeon Hwang et al. [31] used the mixture of thermoplastic (ABS) material and metallic (copper and iron) particles as composite filament for the FDM process and studied the effect on thermo-mechanical properties like tensile strength and thermal conductivity of filament by varying the percent loading of metal particles.

The temperature and fill density were also varied to observe the effects on the tensile strength of the final product. It has been concluded that the tensile strength of the composite material decreases and the thermal conductivity increases by increasing the percent loading of metal particles. The main results were obtained for the thermal conductivity. The coefficient of the specimen's thermal expansion was decreased to 29.5 percent of pure ABS in addition to 50 wt percent of copper particles into the ABS. This showed that the thermal conductivity was increased with the addition of copper particles. An increase in the thermal conductivity of the final product can reduce the problem of distortion of large-scale objects. Hence the composite filament of the ABS and copper particle has the ability to reduce or, to some extent, eliminate the distortion of the final component.

Studies were also carried out on improving the electrical conductivity characteristics of the polymer. Wai Kwok et al. [32] added the carbon black conductive filler material in the polymeric matrix of polypropylene and made filament using a single screw extruder. The compositional analysis was done, which showed that, with an increase in the weight percentage of conductive carbon black filler, the electrical resistivity of polypropylene composite filament decreases. Figure 10.1 shows the variation of resistance with an increasing amount of conductive filler material.

The conclusion was drawn that low resistivity can be achieved with the high percentage loading of the carbon black filler (≥ 30 percent by weight). Using recycled polymer material instead of conventional material available (like ABS, PLA, etc.) as feedstock material for FDM is the right approach towards sustainable and green manufacturing. K.S. Boparai et al. used the recycled Nylon 6 polymer material and composite was made by adding nanofillers in the polymer matrix. The thermal characterization of Nylon 6

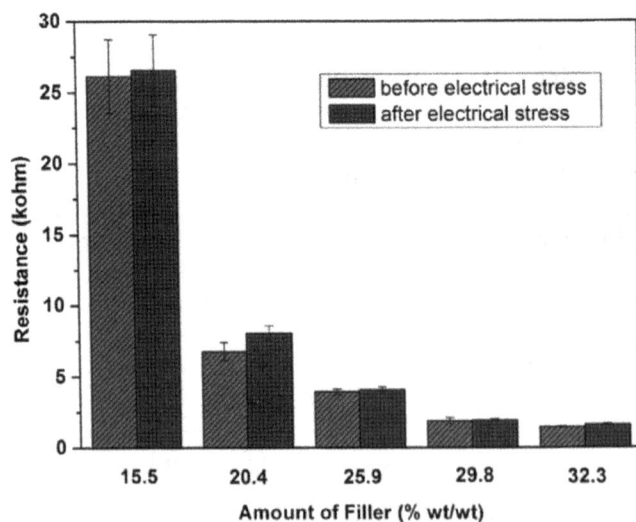

Figure 10.1 Plot of electrical resistance vs amount of conductive filler added. Reprinted from open access article [32]. Copyright, 2017, Elsevier

based nano-composite (NC) the material performed by thermogravimetric analysis (TGA) showed an increase in retardancy level and thermal stability of NC material on the addition of nanofillers in the Nylon 6 polymer. Differential scanning calorimeter (DSC) test showed that with the addition of nanofillers the crystallinity decreases but added filler particles act as a thermodynamic sink improving its thermal stability [33].

10.5.2 Metals

Metal additive manufacturing techniques are rapidly evolving in recent times and found their major application in the aerospace, automobile, tooling, and biomedical sectors [34]. The advantages like in-house production, design, and manufacturing complex shape and lightweight components, reduced waste of material and manufacturing assemblies directly are potential driving components for shifting from conventional techniques to metal AM techniques. Generating complex metal parts with AM offers an economic advantage over the traditional manufacturing process because of lower cost and reduced material wastage [15].

Powder bed fusion (laser beam melting (LBM), electron beam melting (EBM), selective laser sintering (SLS), binder jetting, and sheet lamination are the techniques that are mainly used for metal additive manufacturing [35, 36]. Despite the developments happening in the field of metal AM, there are some more significant challenges that need to be overcome. The properties of the metal components purely depend on the microstructure of the component; hence the main parameters, which are porosity and microstructure of the component, need to be controlled to get the good properties of the component [37]. Also, as the component is printed layer by layer; hence the anisotropy in the properties is observed in the build direction [38].

Using these metal AM techniques for localized and mass production needs a supply of feedstock material. For the material's sustainability, many attempts were made by the researchers to find out the potential of recycling metal waste and utilizing it as a feedstock material for the 3D printing technique [39–41].

The cost of the material is always one drawback of the AM technique; hence recycling the material and using it as feedstock for the AM technique will definitely help to reduce the cost of 3D printed components. Cacace S. et al. [40] performed a comparative analysis of the properties of the component made from recycled powder and standard powder of stainless steel. Both recycled and standard powder were used to manufacture the 3D part using two AM techniques, namely selective laser melting (SLM) and laser metal deposition (LMD). The same process parameters were used for both the powders and the 3D components were printed. The investigation of the properties revealed that small increases in the tensile strength for the recycled powder and mechanical properties were comparable to the standard powder specimen hence it showed that recycled gas atomized powder can be used for AM and it is a sustainable process.

Blake Fullenwider et al. [41] made powders from recycled milling waste chips using mechanical milling and the potential of using that powder in AM was proved by depositing the single tracks and multiple layers using laser engineered net shaping. Stainless steel chips of length 5–20 mm were selected and were ground to powder using two different balls of size 20 mm and 6 mm. The near spherical morphology particles with size around 38–150 μm were obtained. The observation was made that a ball of 20 mm diameter effectively reduces the particle to spherical morphology compared to a 6 mm diameter ball. Also, the powder which was obtained using the mechanical ball milling process had higher hardness than the gas atomization method of producing the powder and this was confirmed by the nanoindentation testing method.

Recycling the metal chip waste by gas atomization (GA) process involves high energy consumption and high cost. Recycling via a newly developed solid-state process that is additive friction stir deposition (AFS-D) is a good alternative for metal chips recycling [42]. In the AFS-D process, chips undergo solid state shear induced plastic deformation, and the bonding takes place between the frictionally heated material and previously deposited layer (Substrate). J.B. Jordon et al. presented the work on directly using the metal chips for the solid state AFS-D process and depositing the fully dense layer with refined microstructure. The aluminum alloy 5083 (AA5083) machine chips were used for the examination. The deposition is done by a K2 MELD machine and an auger feed system was used as a chip feeding system. The microstructural examination showed that a fine equiaxed structure was developed and wrought material type mechanical properties were obtained [43].

A. Montelione et al. [44] performed the characterization of the extra titanium powder which was not sintered and evaluated the changes in the properties obtained when again used in electron beam melting additive manufacturing (EBM AM). The results showed that a change in the microstructure and mechanical properties of individual powder was observed. The powder particle gets contaminated with the oxygen on reusing the powder over multiple build cycles. Hence the work done on recycling the metal waste shows that, with good control over the process, recycling the extra metal powder is a potential option for the cost-effective AM techniques. In the area of recycling industrial metal waste, more research needs to be done.

10.6 DISCUSSION

AM techniques are gaining importance with technological developments in the manufacturing sectors. Having many advantages over conventional techniques like reduced lead time and tooling cost, more flexibility in the manufacturing, less material waste, the capability of producing complex parts at a lower cost compared to conventional process gave the boost to the AM to use in a wide range of sectors. The aerospace and automobile sectors use the AM to fabricate lightweight and customized components. Researchers and scientists can replicate their ideas and innovation in the real world. AM techniques are finding their application in many other sectors like biomedical, rapid tooling, architecture, the toy industry and in day to day used products manufacturing. Polymer and metal are most popularly used as feedstock materials in AM. The cost of the material is the main contributing factor for the overall cost of the 3D printed component. Optimum use of the material is an important parameter for sustainable manufacturing. Researchers have done work on recycling the industrial polymer and metal waste and using them as feedstock material for the AM technique. This chapter demonstrated different material processes used for recycling and the properties obtained.

10.7 CONCLUSION

Additive manufacturing is the future of the manufacturing sector because of the advantages it offers. For sustainable AM, optimum and efficient use of the material is crucial. Also, to make the AM cost-effective, finding out the ways to minimize material cost is one of the main parameters and more research needs to be done in this area. Many attempts are made by the researcher to recycle the polymer and metal waste/scrap and utilize them as feedstock material for the AM techniques. The results obtained for the mechanical properties, flow properties, as well as thermal properties show that it is a potential option and can be good for cost-effective and sustainable AM manufacturing. More work is done on the recycling of polymer waste compared to recycling metal scrap. Finding out more efficient and cost-effective ways for the recycling of polymer and metal can be the key area for the research.

ACKNOWLEDGMENT

We would like to thank the Department of Science and Technology (DST), India for support under the Scheme for Young Scientists & Technologist (SYST) (File No: SP/YO/2019/1240).

ABBREVIATIONS

AM   Additive manufacturing
FDM   Fused deposition modeling
ABS   Acrylonitrile butadiene styrene
PLA   Polylactic acid

REFERENCES

1. ASTM Standard F2792, *Standard Terminology for Additive Manufacturing Technologies.* ASTM International, West Conshohocken, Pennsylvania, 2012.
2. Gebhardt, A., Kessler, J., Thurn, L., *Applications of Additive Manufacturing.* 3D Printing, 101–136. doi: 10.3139/9781569907030.004.
3. Travitzky, N., Bonet, A., Dermeik, B., Fey, T., Filbert-Demut, I., Schlier, L., Schlordt, T., Greil, P., Additive manufacturing of ceramic-based materials. *Advanced Engineering Materials,* 16(6), 729–754. doi:10.1002/adem.201400097.
4. Singh, S., Ramakrishna, S., Singh, R., Material issues in additive manufacturing: A review. *Journal of Manufacturing Processes,* 25, 185–200. doi:10.1016/j.jmapro.2016.11.006.
5. Thomas, D., Costs, benefits, and adoption of additive manufacturing: A supply chain perspective. *The International Journal of Advanced Manufacturing Technology,* 85(5–8), 1857–1876. doi:10.1007/s00170-015-7973-6.
6. Raheem, D., Application of plastics and paper as food packaging materials - An overview. *Emirates Journal of Food and Agriculture,* 25(3), 177. doi:10.9755/ejfa.v25i3.11509.
7. Alabi, O. A., Ologbonjaye, K. I., Awosolu, O., Alalade, O. E., Public and environmental health effects of plastic wastes disposal: A review. *Toxicol Risk Assess,* 5(1), 2019. doi:10.23937/2572-4061.1510021.
8. Gug, J., Cacciola, D., Sobkowicz, M. J., Processing and properties of a solid energy fuel from municipal solid waste (MSW) and recycled plastics. *Waste Management,* 35, 283–292. doi:10.1016/j.wasman.2014.09.031.
9. Wang, Y., Li, Y., Wang, W., Lv, L., Li, C., Zhang, J., Recycled polycarbonate/acrylonitrile-butadiene-styrene reinforced and toughened through chemical compatibilization. *Journal of Applied Polymer Science,* 47537. doi:10.1002/app.47537.
10. Goodship, V., Plastic recycling. *Science Progress,* 90(4), 245–268. doi:10.3184/003685007x228748.
11. Ragaert, K., Delva, L., Van Geem, K., Mechanical and chemical recycling of solid plastic waste. *Waste Management,* 69, 24–58. doi:10.1016/j.wasman.2017.07.044.
12. Liu, S., & Shin, Y. C., Additive manufacturing of Ti6Al4V alloy: A review. *Materials & Design,* 164, 107552. doi:10.1016/j.matdes.2018.107552.
13. Singh, R., & Singh, S., *Additive Manufacturing: An Overview. Reference Module in Materials Science and Materials Engineering.* doi:10.1016/b978-0-12-803581-8.04165-5.
14. Gokhare, V. G., Dr. Raut D. N., Dr. Shinde, D. K., "A review paper on 3D-printing aspects and various processes used in the 3D-printing, *International Journal of Engineering Research & Technology (IJERT),* 6, ISSN: 2278-0181, 2017. 953–958.
15. Ngo, T. D., Kashani, A., Imbalzano, G., Nguyen, K. T. Q., Hui, D., Additive manufacturing (3D printing): A review of materials, methods, applications and challenges. *Composites Part B: Engineering,* 143, 172–196. doi:10.1016/j.compositesb.2018.02.012.

16. Zocca, A., Colombo, P., Gomes, C. M., & Günster, J., Additive manufacturing of ceramics: Issues, potentialities, and opportunities. *Journal of the American Ceramic Society*, 98(7), 1983–2001. doi:10.1111/jace.13700.
17. Wang, X., Jiang, M., Zhou, Z., Gou, J., Hui, D. (2017). 3D printing of polymer matrix composites: A review and prospective. *Composites Part B: Engineering*, 110, 442–458. doi:10.1016/j.compositesb.2016.11.034.
18. Polymer properties database, (Web reference). http://polymerdatabase.com/Plastic%20Products/PlasticToys.html.
19. Nur-A-Tomal, M. S., Pahlevani, F., Sahajwalla, V., Direct transformation of waste children's toys to high quality products using 3D printing: A waste-to-wealth and sustainable approach. *Journal of Cleaner Production*, 122188. doi:10.1016/j.jclepro.2020.122188.
20. Ngo, T. D., Kashani, A., Imbalzano, G., Nguyen, K. T. Q., Hui, D., Additive manufacturing (3D printing): A review of materials, methods, applications and challenges. *Composites Part B: Engineering*, 143, 172–196. doi:10.1016/j.composite-sb.2018.02.012.
21. Mohammed, M. I., Das, A., Gomez-Kervin, E., Wilson, D., Gibson, I., EcoPrinting: Investigating the use of 100% recycled Acrylonitrile Butadiene Styrene (ABS) for additive manufacturing. in 2017 International Solid Freeform Fabrication Symposium. University of Texas at Austin. ttps://hdl.handle.net/2152/89859. .
22. Jiménez L, Mena MJ, Prendiz J, Salas L, Vega-Baudrit J., Polylactic acid (PLA) as a bioplastic and its possible applications in the food industry. *J Food Sci Nut 5*: 048.
23. Auras, R., *Poly(lactic acid). Encyclopedia of Polymer Science and Technology.* doi:10.1002/0471440264.pst275.
24. Murariu, M., Dechief, A.-L., Ramy-Ratiarison, R., Paint, Y., Raquez, J.-M., Dubois, P., Recent advances in production of poly(lactic acid) (PLA) nanocomposites: A versatile method to tune crystallization properties of PLA. *Nanocomposites*, 1(2), 71–82. doi:10.1179/2055033214y.0000000008.
25. Cruz Sanchez, F. A., Boudaoud, H., Hoppe, S., Camargo, M., Polymer recycling in an open-source additive manufacturing context: Mechanical issues. *Additive Manufacturing*, 17, 87–105. doi:10.1016/j.addma.2017.05.013.
26. Tanney, D., Meisel, N. A., & Moore, J. (2017). Investigating material degradation through the recycling of PLA in additively manufactured parts. In *2017 International Solid Freeform Fabrication Symposium.* University of Texas at Austin. https://hdl.handle.net/2152/89857.
27. Zenkiewicz, M., Richert, J., Rytlewski, P., Moraczewski, K., Stepczyńska, M., Karasiewicz, T., Characterisation of multi-extruded poly(lactic acid). *Polymer Testing*, 28(4), 412–418. doi:10.1016/j.polymertesting.2009.01.012.
28. Zander, N. E., Park, J. H., Boelter, Z. R., Gillan, M. A., Recycled cellulose polypropylene composite feedstocks for material extrusion additive manufacturing, published in American chemical society OMEGA. doi: 10.1021/acsomega.9b01564.
29. Ito, H., Hattori, H., Okamoto, T., Takatani, M., Thermal expansion of high filler content cellulose-plastic composites. *Journal of Wood Chemistry and Technology*, 30(4), 360–372. doi:10.1080/02773810903537119.

30. Zander, N. E., Park, J. H., Boelter, Z. R., Gillan, M. A., Recycled cellulose polypropylene composite feedstocks for material extrusion additive manufacturing, published in American chemical society OMEGA. doi: 10.1021/acsomega.9b01564.

31. Hwang, S., Reyes, E. I., Moon, K., Rumpf, R. C., Kim, N. S., Thermomechanical characterization of metal/polymer composite filaments and printing parameter study for fused deposition modeling in the 3D printing process. *Journal of Electronic Materials*, 44(3), 771–777. doi:10.1007/s11664-014-3425-6.

32. Kwok, S. W., Goh, K. H. H., Tan, Z. D., Tan, S. T. M., Tjiu, W. W., Soh, J. Y., Glenn-Ng, Z. J., Chan, Y. Z., Hui, H. K., Goh, K. E. J., Electrically conductive filament for 3D-printed circuits and sensors. *Applied Materials Today*, 9, 167–175. doi:10.1016/j.apmt.2017.07.001.

33. Boparai, K. S., Singh, R., Fabbrocino, F., Fraternali, F., Thermal characterization of recycled polymer for additive manufacturing applications. *Composites Part B: Engineering*, 106, 42–47. doi:10.1016/j.compositesb.2016.09.009.

34. Yakout, M., Elbestawi, M. A., Veldhuis, S. C., A review of metal additive manufacturing technologies. *Solid State Phenomena*, 278, 1–14. doi:10.4028/www.scientific.net/ssp.278.1.

35. Frazier, W. E., Metal additive manufacturing: A review. *Journal of Materials Engineering and Performance*, 23(6), 1917–1928. doi:10.1007/s11665-014-0958-z.

36. Herzog D, Seyda V, Wycisk E, Emmelmann C., Additive manufacturing of metals. *Acta Materialia*, 117(3), 371–392. doi:10.1016/j.actamat.2016.07.019.

37. Maskery, I., Aboulkhair, N. T., Corfield, M. R., Tuck, C., Clare, A. T., Leach, R. K., ... Hague, R. J. M., Quantification and characterisation of porosity in selectively laser melted Al–Si10–Mg using X-ray computed tomography. *Materials Characterization*, 111, 193–204. doi:10.1016/j.matchar.2015.12.001.

38. Somireddy, M., Czekanski, A., Anisotropic material behavior of 3D printed composite structures – Material extrusion additive manufacturing. *Materials & Design*, 108953. doi:10.1016/j.matdes.2020.108953.

39. Ford, S., Despeisse, M., Additive manufacturing and sustainability: An exploratory study of the advantages and challenges. *Journal of Cleaner Production*, 137, 1573–1587. doi:10.1016/j.jclepro.2016.04.150.

40. Cacace, S., Furlan, V., Sorci, R., Semeraro, Q., & Boccadoro, M., Using recycled material to produce gas-atomized metal powders for additive manufacturing processes. *Journal of Cleaner Production*, 122218. doi:10.1016/j.jclepro.2020.122218.

41. Fullenwider, B., Kiani, P., Schoenung, J. M., Ma, K., From recycled machining waste to useful powders for metal additive manufacturing. *SpringerBriefs in Applied Sciences and Technology*, 3–7. doi:10.1007/978-3-030-10386-6_1.

42. Griffiths, R. J., Perry, M. E. J., Sietins, J. M., Zhu, Y., Hardwick, N., Cox, C. D., Rauch, H. A., Yu, H. Z., A perspective on solid-state additive manufacturing of aluminum matrix composites using MELD. *Journal of Materials Engineering and Performance*. doi:10.1007/s11665-018-3649-3.

43. Jordon, J. B., Allison, P. G., Phillips, B. J., Avery, D. Z., Kinser, R. P., Brewer, L. N., Cox, C., Doherty, K., Direct recycling of machine chips through a novel solid-state additive manufacturing process. *Materials & Design*, 108850. doi:10.1016/j.matdes.2020.108850.

44. Montelione, A., Ghods, S., Schur, R., Wisdom, C., Arola, D., Ramulu, M., Powder reuse in electron beam melting additive manufacturing of Ti6Al4V: Particle microstructure, oxygen content and mechanical properties. *Additive Manufacturing*, 101216, 2020. doi:10.1016/j.addma.2020.101216.

Chapter 11

Tool condition monitoring in mechanical micromachining

Manoj Kumar K., Baburaj Mb, and Nithin Tom Mathewc

11.1 INTRODUCTION

Miniaturization of components leads to high heat transfer rate, low power consumption, space and material optimization, high surface quality, ease in transportation, enhanced quality of life, and health care. Micro products are used in biomedical, information technology, medical and automobile, communication, MEMS, and electronics, etc.

Micromachining is one of the most basic technologies to produce micro-sized products and it bridges the gap between micro electro-mechanical system (MEMS) and conventional machining processes (Taniguchi 1983). The microfeature or micro products are manufactured with the assist of various micromanufacturing technologies such as focused ion beam, laser machining, mechanical micromachining, etching, micro-electro discharge grinding, LIGA, ultrasonic vibration assisted micromachining, electroplating, and micro stereolithography, etc. (Chae et al. 2006; Robinson and Jackson 2005; Dornfeld et al. 2006; Rajurkar et al. 2006; Qin et al. 2010a, 2010b).

Among these methods, mechanical micromachining can machine a wide variety of materials with superior material removal rate, finish, and low cost. Mechanical micromachining can deliver high aspect ratio three-dimensional micro components with greater geometric complexity. Mechanical micromachining is defined as the mechanical cutting of features with geometrically defined cutting edge(s) of size less than 1 mm and with no restrictions on the size of the components (Masuzawa and Tonshoff 1997; Dornfeld et al. 2006; Jain 2010). The various types of mechanical micromachining process are micro milling, micro turning, and micro drilling (Chae et al. 2006; Rahman et al. 2006; Asad et al. 2007; Liow 2009).

One of the major challenges in micromachining is to identify various methods to optimize the machining process. The process of optimization includes optimization of micromachining parameters, reducing the operational cost, increasing the product output, and minimizing the tool wear. Numerous methods are implemented to address these factors. In mechanical micromachining, tool wear has a significant role in the success of the process. The micro nature of the tool makes the investigation more complex.

DOI: 10.1201/9781003246466-11

Due to the difficulty involved in this process, it is challenging to identify the tool wear in the microscale. The worn-out tool can influence the condition of the part produced. Therefore, the worn-out tool must be changed at a suitable time during machining.

Automated manufacturing systems have become a part of Industry 4.0. Artificial intelligence, big data analysis, the Internet of Things, and high-performance computer systems together make the interconnectivity of machines and also make the condition monitoring in manufacturing much easier. In-situ and remote monitoring of machining processes and machine tools using sensors have become an important part of a fully automated manufacturing system. When it comes to micro manufacturing, monitoring of tool condition becomes more important than the machine tool condition.

The micro tool is subjected to comparatively large vibration and cutting force during the micromachining as the tool wear increases. In order to retain precision as well as quality in the components, it is necessary to monitor the condition of the micro tool. Hence, there is a requirement to build a tool condition monitoring (TCM) system for the mechanical micromachining process to suggest to the operator about the condition of the micro tool. Researchers have created TCM systems using various sensors such as touch sensor, power sensor, acoustic emission (AE) sensor, vibration sensor, torque sensor, vision sensor, and temperature sensor, etc., in different machining processes. The application of various sensors comes with certain objectives. A few of them are used to monitor the machine to perform diagnostics. Some other sensors are used to observe the condition of the tool, some for investigating the work material, and others to monitor the chip formation, energy consumption, variation in temperature, etc. Among these sensors, it is observed that AE, accelerometer, and cutting force sensors are commonly used for examining the micro tool condition (Dan and Mathew 1990; Byrne et al. 1995; Yeo et al. 2000; Chelladurai et al. 2008).

11.1.1 Micromachining

Micromachining is defined by various researchers in numerous ways depending upon the feature size, industry, and focus of interest. In mechanical micromachining, the interaction between a sharp solid cutting tool with the workpiece material leads to the material removal in the form of chips (Alting et al. 2003; Dornfeld et al. 2006). Micromachining is classified broadly under two major categories, namely mask based and tool based.

Mask based technologies are employed for obtaining finer precision, especially at atomic levels. They are mainly applied for semiconductor processing in which silicon materials undergo chemical and dry processes. High precision micro components can be produced by these processes. However, they have limitations such as high initial cost, low material removal rate, and limited material selection. Moreover, they are mostly used to manufacture two dimensional shapes. Mask based micromachining is also known as MEMS based micromachining methods. The early fabrication of these micro components was mainly focused on employing conventional semiconductor techniques to develop MEMS devices and other microelectronic products and hence they are called MEMS micro manufacturing or lithography based micro manufacturing. The most popular techniques in MEMS based micro manufacturing include chemical etching, photolithography, plating, and LIGA, etc., and imprint lithography, and soft lithography. The initial application of MEMS based micromachining techniques was extensively for semiconductor and chromatography application. Unfortunately, MEMS based micromachining processes have many restrictions including (1) the necessity of many repetitive processing steps to generate 3D structures, (2) the inability to generate continuously curved structures, (3) increased times, (4) increased production costs, (5) low production rates, (6) material restriction, and (7) limited aspect ratio.

The tool based or mechanical micromachining is the machining of features with geometrically defined cutting edge/edges of size less than 1 mm and with no restrictions on the size of the part. Jain (2010) defined micromachining as the machining processes with material removal in the order of micro/nano with no restriction on the size of the components. Tool based micromachining techniques use solid tools and do not require any expensive setup like mask based processes and hence they can produce micro components at low cost. They can create three dimensional shapes in most of the materials such as metallic alloys, composites, polymers, ceramic materials, etc., with higher material removal rate (Masuzawa and Tonshoff 1997; McGeough 2002; Chae et al. 2006; Rajurkar, et al. 2006; Asad et al. 2007; Jackson 2007; Liow 2009). Tool based micromachining helps to bridge the gap between macro and nano domains for making functional components (Chae et al. 2006).

The main differences between tool based micromachining and macromachining are mainly in the machining mechanism. In conventional macromachining, the material removal happens primarily due to shearing which results in the chip formation. The cutting tools can completely remove material from the surface of the workpiece and produce chips as the depth of cut or undeformed chip thickness is generally more than that of the cutting tool edge radius. In the traditional machining process, it is found that the ratio of the undeformed chip thickness to the tool edge radius is large. In the case of mechanical micromachining, the material removal depends on the size effect and the minimum uncut chip thickness.

In mechanical micromachining, the generation of chips can be attained only if the undeformed chip thickness reaches a critical value. When the uncut chip thickness is less than a critical minimum chip thickness, the material will be exposed to compression by the action of the micro tool and then it will recover back (elastic deformation) after the tool passes.

Due to this there is no material likely to be removed and as a result ploughing occurs rather than shearing. When the uncut chip thickness reaches the minimum chip thickness the chips are formed due to shearing of the workpiece, however some elastic deformation still happens. However, if the uncut chip thickness increases beyond the minimum chip thickness the elastic deformation phenomena reduce considerably leading to shearing of the material and results in the generation of chips.

Kim and Kim (1995) analytically studied the variations in the cutting mechanism between the macro-machining and micromachining processes. Koc and Ozel (2011) observed that the minimum chip thickness is found to be between 5 percent and 38 percent of the edge radius in tool based micromachining processes for different materials. The different types of mechanical micromachining processes are micro milling, micro turning, and micro drilling, etc.

The micro turning process can manufacture micro cylindrical products like micro pins, shafts, and electrodes. Micro drilling is a process to make micro holes, for example, micro nozzles, vacuum relief holes, flow control restrictions, vent holes, etc. Micro milling is used to remove the material from the workpiece to create shapes like micro channels, 2½D profiles, and 3D free form surfaces. Out of these conventional micro cutting processes, end milling is universally used for manufacturing micro parts because of the versatility in process, ease in making complex geometries with intricate contours, improved surface finish, high dimensional accuracy, and having the ability to machine difficult to machine materials.

11.2 TOOL CONDITION MONITORING (TCM)

Manufacturing industries are undergoing a dramatic change. The present trend is towards an adaptive control system for cost saving, reducing production time, and improving the part quality. A sensor based manufacturing is the backbone of this evaluation for adaptive control. The development of advanced sensors and computational facilities for signal processing permit improved information gathering enabling process optimization and control. The adaptive control includes the in-process quality control, machine tool diagnostics, tool condition detection, etc.

The cutting tool is an integral part of the machining process, which can affect the surface quality of the product if it is not in good condition. The major goal of manufacturing engineering is to improve productivity and quality of product with low machining time and labour cost. This led to the development of in-process TCM systems. The manual process of monitoring the tool is difficult, time consuming, and often can mislead. However, the development of sensor technologies helps to develop the TCM system, which makes the machine tool more intelligent and adaptive during the machining process. Tool condition monitoring is regarded as critical for the following reasons:

- Unmanned production is feasible only if the tool breakage and wear can be monitored.
- Tool wear affects the surface quality of the part fabricated.
- The present tool change is made based on prediction of tool life which does not consider unexpected failures. This leads to an unnecessarily high number of changes because the full lifetime of tools is not considered and, consequently, valuable production time is lost.
- Automated production is not possible without wear monitoring.

TCM improves the reliability and promotes automation of the manufacturing process. One of the major activities in TCM is to detect tool failure which includes tool wear, cracking, chipping, and fracture of the tool. The in-process TCM system detects any malfunctions during the process and indicates to the operator about the status of the cutting tool (Teti 1992; Tonshoff and Inasaki 2001; Wang and Gao 2006). Apart from the tool condition monitoring with the in-process system, it also helps in understanding the cutting process for optimization. Apart from these measurements, a real time recognition of the machining process is also essential for improved insight of the cutting phenomena. Literature indicates that various sensors have been used in the development of TCM systems.

Sensors are devices that detect a change in a physical stimulus and turn it into a signal which can be recorded (Tonshoff and Inasaki 2001; Dornfeld and Lee 2008). The sensors used for TCM are classified as direct and indirect sensors. In the direct measuring system, the actual quantity of the measured variable such as tool wear is measured directly, whereas the indirect method employs suitable auxiliary quantities such as cutting force, acoustic emission (AE), vibration, etc., for extracting the actual quantity. The direct measuring system has higher accuracy, whereas the indirect method is less complex and more suitable for practical applications. Direct measurement can be carried out only after some time intervals, i.e., the machining must be stopped which results in loss of time and subsequently results in high cost. The indirect method helps in detecting all the changes in the machining process continuously (online) and responds quickly to conditions such as tool breakage, fracture, and tool wear status, etc. The tool damage can be prevented by using the indirect method of measurement whereas with the direct measurement the tool damage cannot be detected until the machining cycle is over.

Tool monitoring is an important concern in micromachining. The major issues observed in micromachining related to micro tools are tool wear, deflection, and run out. Researchers are using different methods to monitor the condition of micro tools. Some of the methods used to monitor micro tools are based on cutting force pattern based (Jemielniak and Arrazola 2008), acoustic emission (, Jemielniak and Arrazola 2008), neural network (Tansel et al. 2000), genetic algorithm, laser system, and scanning electron microscopic imaging.

One of the major issues found in micro milling is the unpredictable nature of tools. Tool failure due to build up edge, cutting edge chipping off, and increase of cutting-edge radius due to wear reversely affect the quality of machined surface. Tool-workpiece interface friction is also a reason for tool wear when cutting with very high feed values. Elastic recovery during micro cutting also causes flank wear and decrease in flank wear with increase in workpiece inclination angle observed in micro ball end milling (Arif et al. 2012). It is possible to identify the presence of tool wear from force pattern (Rahman et al. 2006). They observed an increasing trend in cutting force in consecutive cutting cycles. Few studies are reported on estimation of tool wear based on cutting force for metals and non-metals (Tansel et al. 2000). Other than these, geometrical variations on the cutting edge also cause dimensional change in machined features.

The diameter of the tool used in micro end milling is very small. Small diameter tools tend to deflect at high feed rate (Chae et al. 2005, Ozel and Liu 2009) and this may cause tool failure. The combined effect of tool deflection and geometrical variations in the cutting-edge during machining is adversely affecting the accuracy of the machined feature (Escolle et al. 2015). Few studies are reported on the effect of tool deflection in flat end milling (Uriarte et al. 2008, Tansel et al. 1998 Rodriguez and Labarga 2013) and ball end milling (Dow et al. 2004). The effect of tool deflection can be reduced by keeping the overhanging length as small as possible .

Tool runout is also considered as an important parameter in micro end milling because of the reduction in size. The main causes for run out are the problems in tool holder, tool shank, and spindle (Perez et al. 2007, Bao and Tansel 2000a). Along with the above-mentioned factors, tool deflection also causes runout (Kang and Ahn 2007). The presence of runout offers more load on to a few teeth and less on others which results in low surface finish and high tool wear. Uncut chip thickness is very much sensitive to runout due to radius variations. The presence of runout changes the trajectory of cutting edge from the expected (Li et al. 2007), and the existence of runout can be easily identified from the peak force variations and surface topography. Cutter runout is used to express in terms of runout length and angle. Few attempts are made to study runout effect in micro flat end mill (Rodriguez and Labarga 2013,) as well as in ball end milling (Xu et al. 2015). Researchers used different available methods to measure runout, such as microscopic (Jin and Altintas 2012), analytical run out estimation from cutting force data (Bao and Tansel 2000b), and incremental position sensor (Wojciechowski 2015).

The major direct type of sensors used for TCM systems are proximity sensor, radioactive sensor, and vision etc. The indirect type of sensors consist of force, AE, power, and vibration, etc. (Byrne et al. 1995; Tonshoff and Inasaki 2001). Among the sensors, force, AE, and vibration were widely used for monitoring the condition of the tool in mechanical micromachining.

11.2.1 Acoustic emission (AE) sensor

AE is low amplitude and high frequency elastic waves produced due to a quick release of strain energy from localized source/sources within the material (Ballantine et al. 1997; Tonshoff and Inasaki 2001). The waves generated travel to the surface, where they can be sensed by a piezoelectric transducer. During a conventional machining process, AE signals are generated due to plastic flow of material or fracture. The frequency range of the AE signal (100 kHz to 2 MHz) is much larger than that of the machine vibrations and environmental noises and hence they can detect micro scale deformations even in noisy environments without interfering with the cutting operation. Hence, they are ideal for characterizing the material removal activity in macro and micro scale. AE based monitoring systems have been available for approximately 30 years.

The major sources of AE during the metal removal process are the following (Dan and Mathew 1990; Teti 1992; Byrne et al. 1995; Xiaozhi and Beizhi 2007).

  (i) Plastic deformation during the machining in the work material and chip.
 (ii) Friction between the flank face of the tool and the work material leading to flank wear.
(iii) Friction between the chip and the rake face of the tool leading to crater wear.
 (iv) Collisions between tool and chip.
  (v) Chip breakage during the machining process.
 (vi) Tool fracture during the machining process.

The AE signals are mainly classified under continuous and transient signals. The continuous signal is associated with the deformation of material and friction (i)–(iii) whereas the transient signal consists of burst signal superimposed on the continuous signal which is associated with the chip and tool breakage (iv)–(vi) (Xiaoli 2002). When compared to other sensing methods like cutting force, AE provides better sensitivity to deformation mechanisms in ultra-precision machining Dornfeld 2006).

The amplitude of AE energy is a significant factor considered to identify various sources during the material removal. The amplitude of AE energy due to the rapid transmission of a crack is much higher than the amplitude of AE energy immediately prior to crack propagation. During the generation of continuous chips, low level continuous AE signals were generated from the machining zone. The initiation of fracture leads to a high-amplitude AE burst signal suggesting chip breaking. In mechanical micromachining, the difference in AE energies is employed to observe the brittle fracture. Acoustic emission has been found to be sensitive to minor variations in grinding regime.

The main AE parameters considered are root mean square value (RMS), peak to peak value, ringdown counts, rise time, and dominant frequency. Dan and Mathew (1990) reviewed the application of AE sensor for tool wear monitoring in turning. Generally, the AE signals were correlated with the tool wear. The gradual increase of the AE indicates the growth in the flank wear. The frequency spectrum of the tool is found to be between 100 and 250 kHz. A sudden increase in amplitude of AE signal was detected during tool fracture. The burst type of AE signals is mostly produced at the time of tool breakage. Large amplitude AE signals were also observed with tool cracking, chipping, and fracture.

11.2.2 AE signal analysis

AE signals are non-stationary and often consist of overlapping transients, whose waveforms are unspecified. The major challenge in AE signal processing is to isolate physical parameters, tool wear for example. The signal consists of both frequency and time and various processing methods are to be adopted to analyse the AE signals. The various signal analysis approaches in AE include parameter analysis, time series analysis, and Fourier transform.

Parameter analysis is capable of recording and dealing with large amounts of data and analysing them in real time, thus making it the main method for AE signal analysis. The parameters include AE event, rise time, ring down count, RMS, peak amplitude, etc., based on the AE waveform. AE event is the micro-structural movement that generates elastic waves in a material under load/stress. Ring down count is the number of times the signal amplitude exceeds the present reference threshold. RMS voltage measures the off-signal intensity. To get the accurate analysis, the relationship between the AE sources and the signals is to be identified properly which otherwise can lead to error or inadequate results. For the experiments, it was clear that with the increase in tool wear there is a notable increase in the AE signals during the metal removal processes. Adaptive time series modelling along with band filtered energy analysis is adopted in conjunction with the cutting condition. In Fourier analysis the AE energy signal was decomposed to Fourier transform to extract the information from the signal. The AE data is cut to different frequency components.

II.2.3 Accelerometer sensor

Accelerometers are devices commonly used for sensing vibrations or different modes of acceleration generated in a body. Mechanical engineers often use them to monitor vibration in aerospace, manufacturing, and automotive applications. The device is mounted directly on the object that vibrates, where the vibration energy gets converted into an electrical signal which is comparative with the momentary acceleration of the object. The vibration data for the analysis purpose at a high sampling rate is ensured by connecting it further to a data acquisition system. Such measurements are important for diagnosing problems associated with machines or structures which are subjected to periodic stresses. Any malfunctioning in the structural part of a machine tool gives rise to an increased vibration level which can affect the performance level of the machine during the course of time. Thus, vibration monitoring in machine tools is highly beneficial for ensuring part quality, tool integrity, and tool life and for reducing unnecessary downtime caused due to unexpected tool failure. Different types of accelerometers are used for industrial applications like piezoelectric, piezoresistive, capacitive, hall effect, magneto resistive, etc., where the working principle decides the accuracy of measurement. MEMS based accelerometers are another category which are normally used in consumer electronics, mobile devices, and in wearables. Its small size helps in mounting it directly on a PCB.

The piezoelectric accelerometer is used in the TCM system to measure the machine vibrations resulting from the oscillations of the cutting forces in range below 15 kHz. It consists of seismic mass and spring-damping systems attached with displacement pick up. The instrument relies on the piezoelectric effect, where a voltage is produced across certain crystals when they are stressed. Piezoelectric transducers are highly sensitive when compared with other types of transducers and it has wide frequency and dynamic ranges, and it exhibits good linearity in the operating ranges. The device is reliable and robust, and it provides stable operation over a long period. It is designed to be used in rough environmental conditions and with relatively low investment (Tonshoff and Inasaki 2001). During the metal cutting process, vibrations are generated when there is any malfunction such as fracture of tool or excessive wear. The generated signals due to the vibration can be observed and the data can be used in numerous ways for TCM systems.

II.2.4 Cutting force sensor

The understanding of cutting force and spindle or drilling torque data is essential in confirming the process optimization involved in machining. The process capability can be increased with the analysis of cutting force before the start of production thereby increasing productivity. The recognition of tool collisions, overloads, and tool breakage can also be observed with the aid of cutting force sensors.

In tool manufacturing and production technology the use of dynamometers, which measure all the components of the cutting force, are invaluable. They are utilized in comparing, evaluating, and choosing materials, tools, and machines. Further areas of use result from evaluating the breakage behaviour of tools, defining optimum cutting conditions, and the chip formation process and their influence on cutting forces. Piezoelectric force measuring systems are significantly different from other methods of measurement. The forces acting on the quartz crystal element are transferred to a proportional electric charge. The measuring path of a piezoelectric force measuring element amounts to just a few thousandths of a millimetre. The measuring range of such an element is very large.

The essential properties required for a good dynamometer are high rigidity, good linearity, small cross sensitivity, high sensitivity, high natural frequency, ease of measurement, and light weight (Sun et al. 1982). There are different methods available to measure cutting force; among which piezoelectric type of force sensors have become more popular due to their compactness and excellent dynamic response (Tacuna Systems 2021). The primary concern in the usage of dynamometers in micromachining force measurement is stiffness and integrity of the machine. The addition of a load cell on the machine bed or other parts of the machine should not affect the stiffness and integrity of the machine. Dynamometers can be classified as stationary and rotating types (Kistler 2014). In drilling and milling the stationary dynamometer is placed directly between the workpiece and machine table or directly on the turret and tool in case of lathe. The captured data using a dynamometer has to be transmitted to the monitor system using insulated wired connections. A rotating dynamometer is normally fitted to the machine spindle and the drill or milling cutters are fixed on the rotating dynamometer and the captured data transferred to the monitor system using near field telemetry technique.

The monitoring system converts the analogue signals into digital form after primary conditioning (Jemielnaik 1999). Charge amplifiers are generally used to convert the generated charge into proportional voltage in the case of piezoelectric dynamometers. Later the system must be calibrated to find the relation between developed charge and force. The ratio of charge and the corresponding force is normally referred to as sensitivity of the dynamometer. The unit of sensitivity of the piezoelectric dynamometer is basically given in pC/N.

The basic principle of piezoelectric force sensors is piezoelectric effect. The generation of electric charge as a response to the applied mechanical strain inside the piezoelectric materials is called piezoelectric effect. The formation of electrical dipoles within the piezoelectric material, depending on the speciality in their crystal structure (non-centrosymmetric), has effectively been used to capture the cutting forces (Soin et al. 2016). The materials such as natural biological piezoelectric materials (bone, tendon, etc.), natural crystals (quartz, Rochelle salt, etc.), synthetic ceramics (zirconium titanate, barium titanate, potassium niobate, PZT, etc.), and synthetic polymers (poly vinylidene fluoride, polyimide, cellular polypropylene, etc.) are some of the examples for piezoelectric material (Soin et al. 2016). Out of these, quartz is mainly used as a sensor material in cutting force dynamometer (Davis 1971). The preloaded washers with different patterns made from quartz are placed in the stainless-steel casing as per their phase angles and polarity. They are used to measure tensile, compressive, and shear forces in the specified directions (Sun et al. 1982).

The dynamometer can be classified as 1-component dynamometer, 2-component dynamometer, and multi component dynamometer based on the number of forces that can be measured (Kistler 2014). The charge developed inside the crystal is very small and an electro-static charge amplifier is connected to the transducer to convert the developed charge to a measurable form. The piezoelectric crystal output varies with respect to the environmental condition, especially temperature, and it is commonly known as drift. Drift can be eliminated up to an extent by earthing the transducer output. To reduce drift the transducer must be properly heat insulated (Davis 1971).

Arrangements of crystals for 3- and 4-component sensors are shown in Figure 11.1. In the 3-component sensor, four sets of crystals are arranged in

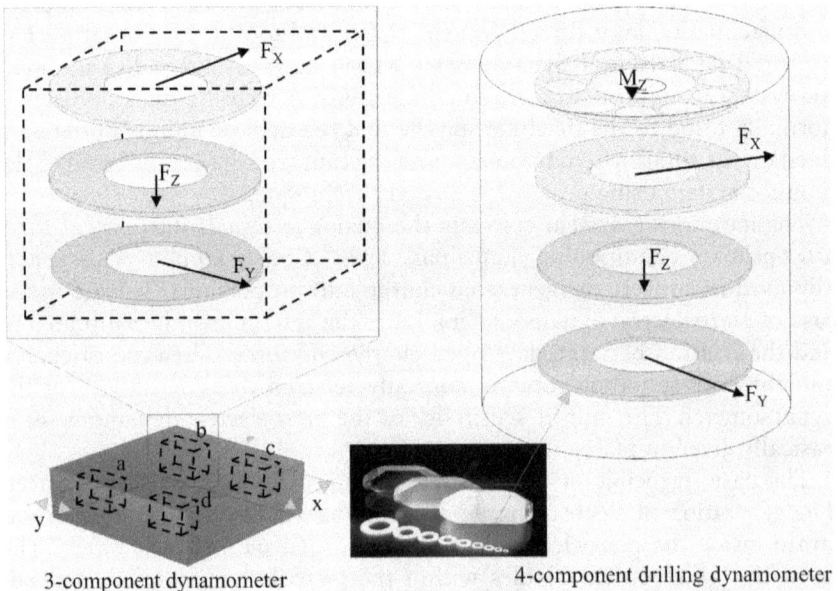

3-component dynamometer   4-component drilling dynamometer

Figure 11.1 Crystal arrangement for 3- and 4- component dynamometer (Kistler 2009)

between preloaded top and bottom plates and each set contains three shear crystals each measuring force in three mutually perpendicular directions. In 3-component sensors, shear sensitive plates are arranged in a circular manner so that their shear sensitive axes become tangential. This type of arrangement helps to measure torque.

In multi-component the forces in Cartesian coordinates are obtained by summing up the corresponding forces from each set as given in Eqn. 11.1 (Kistler 2009).

$$\left.\begin{aligned}
F_X &= F_{Xa+b} + F_{Xc+d} \\
F_Y &= F_{Ya+d} + F_{Yb+c} \\
F_Z &= F_{Za} + F_{Zb} + F_{Zc} + F_{Zd} \\
M_X &= y\left(F_{Za} + F_{Zb} - F_{Zc} - F_{Zd}\right) \\
M_Y &= x\left(-F_{Za} + F_{Zb} + F_{Zc} - F_{Zd}\right) \\
M_Z &= y\left(-F_{Xa+b} + F_{Xc+d}\right) + x\left(F_{Ya+d} - F_{Yb+c}\right)
\end{aligned}\right\} \tag{11.1}$$

It is clear from Eqn. 11.1 that the moments in X-, Y-, and Z-directions are estimated based on the forces and the position of sensors from the centre. The force and torque in micromachining is very low compared to macro-machining. The dynamometer must be calibrated properly to get the actual values. Another solution for this is to use a torque sensor simultaneously with the multicomponent dynamometer (Ravisubramanian and Shunmugam 2015).

11.3 TOOL MONITORING

It is very difficult to detect or identify the tool failures in micro cutting and one of the possible strategies to identify tool failure is the usage of a proper feedback system. Cutting force is one of the effective parameters which can be used in the feedback system especially for milling (Ogedengbe et al. 2011) and torque can be the best parameter for drilling (Botsaris and Tsanakas 2008). The major reasons for tool failure in micro end milling or drilling are the fracture because of excessive cutting force due to improper cutting parameter selection, and tool stiffness, tool run out, and rubbing of elastically recovered workpiece material.

Tool wear, tool deflection, and tool failure are the three parameters which can be identified from cutting force signals. The magnitude of amplitude and instantaneous forces are commonly used for tool conditioning monitoring. Here the instantaneous forces define the shape of the force plot. There are cutting force prediction models developed by the researchers which can be directly used for tool life monitoring since the tool wear and failure are connected with cutting forces (Mohanraj et al. 2020).

In micro milling or drilling the tool must rotate at a very high speed to achieve a reasonably good cutting velocity since the tool diameter is very low. Other than spindle frequency there are two other frequencies that have to be considered for effective force measurement. Those frequencies are tooth pass frequency and sampling frequency. Spindle frequency is the frequency at which the spindle rotates; tooth pass frequency is the frequency at which the cutting edges touch the workpiece; and sampling frequency is the frequency at which the force data is collected. Also, the sampling frequency has to be set at more than three times the tooth pass frequency in order not to miss out the information.

According to Fard and Bordatchev (2013), the measured forces consist of three components such as static, dynamic, and quasi-dynamic. Static data is corresponding to deflection of tools, dynamic data gives information on process dynamics, and quasi-dynamic data provides information on force variation within each cycle (peak to valley) due to tool engagement with the workpiece. Static and quasi-dynamic data can be estimated in terms of mean and peak to valley forces respectively.

In a multi-tooth end mill, the presence of run-out offers more load on a few teeth and less on others. This is because of the variations of effective end mill radius, and it leads to variations in the expected trochoidal cutting edge trajectory (Li et al. 2007). The reasons for end mill run-out are the inaccuracies in the tool holder, tool shank, and spindle (Perez et al. 2007). Other than these, tool deflection and tool vibration due to variable reasons also contribute to additional run-out (Kang and Ahn 2007) especially for long flutes at very high spindle speed. Tool run-out can be easily identified from the force pattern. The effect of uneven loading of cutting-edge results in variation in the amplitude force as shown in Figure 11.2. The condition of tool edges can be monitored by looking into the variation in the amplitude of forces corresponding to the cutting edge.

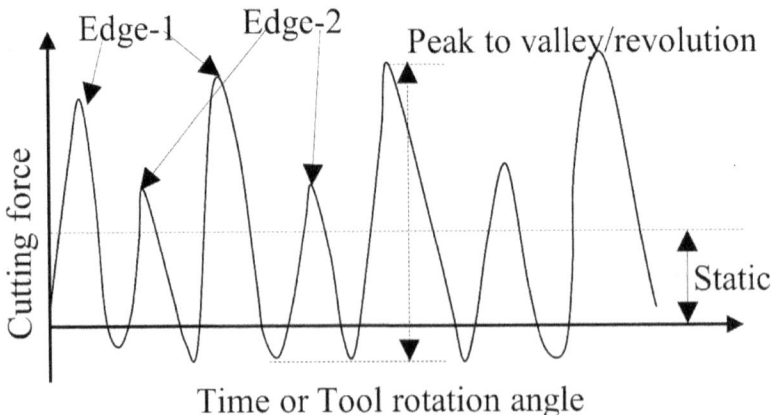

Figure 11.2 Effect of tool runout on cutting force

Fast Fourier transform (FFT) analysis can also be used to assess the machining condition. In the case of stable machining the significant peaks are observed only at frequencies corresponding to spindle and tooth pass frequencies. Significant peaks are also observed even at frequencies corresponding to the multiples of spindle and tooth pass frequencies in unstable machining (Sahoo et al. 2020).

Micromachining cutting force is an excellent indicator of tool wear. It is understood from the existing study that the cutting force increases with tool wear especially when the grain size of the workpiece material is comparable with the cutting-edge radius. But the tool wear in micromachining is unpredictable especially towards the end of tool life (Alhadeff et al. 2020). This is because the higher forces developed due to action of worn cutting edge create an extra tool deflection which ultimately leads to tool failure (Oliaei and Karpat 2016).

There are many tool monitoring methodologies developed based on cutting force signals by various researchers and a few of them are discussed below. A neural network-based tool life estimation method considering force data as input was presented in Tansel et al. (2000) work. The schematic representation of their model is given in Figure 11.3. This model uses feed and thrust force data as input, and they found that the cutting force increases with increase in tool wear for soft materials. Later, Malekian et al. (2009) proposed a methodology to monitor tool condition monitoring using force signals using a neuro fuzzy model. Their model was successful for predicting tool wear for higher feeds and speed conditions. Cutting conditions at low feed and speed conditions are entirely different from the higher value conditions; also the limited bandwidth of the force sensor may be the possible reason for wrong prediction at lower feed-speed values.

Another indirect indication of tool wear is the variation in torque peaks with respect to the machining distance or time. Hong et al. (2016) used wavelet packet transform and Fisher's linear discriminant together for tool monitoring. They experimentally found that the torque peaks are increasing rapidly with respect to cutting time for tools with premature failure and gradually if no premature failure occurred. An empirical statistical method has been proposed by Li and Zhu (2021) for in situ wear prediction in micro

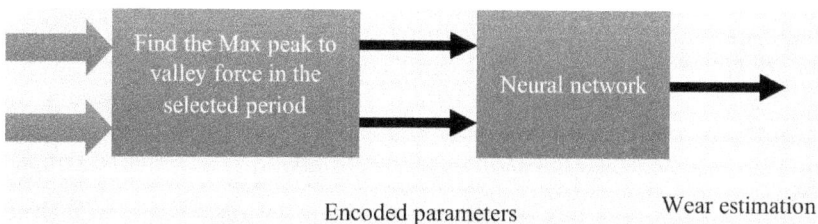

Find the Max peak to valley force in the selected period

Neural network

Encoded parameters   Wear estimation

Figure 11.3 Tool wear estimation model proposed by Tansel et al. (2000)

milling based on the statistical parameters of machining parameters including milling length, wear, and force. They compared the proposed model with the results obtained from the neural network model and experimental result. The proposed model predicts wear with a correlation coefficient of 0.988 and a root mean square error of 127.2. An integrated tool wear forecasting method based on historical data has been demonstrated by Zhang et al. (2022). They used a particle filter algorithm and long short-term memory network together to estimate the tool wear. The cutting force and tool vibration data are used as input to estimate the stochastic tool wear. The estimated tool wear data is considered to predict cutting forces in micro milling along with size effect consideration. The modified model can predict cutting force with an average error of 11.67 percent.

11.4 TEMPERATURE SENSORS

Temperature sensors are basically used for monitoring temperature in various applications like automobiles, scientific and laboratory equipment, domestic appliances, medical applications, HVAC (heating, ventilation, and air conditioning) devices, manufacturing industries, power plants, etc. Commonly used temperature measuring devices are thermistors, thermocouples, RTDs (resistance temperature detector), etc.

Thermistors are a kind of semiconducting material which have resistance higher than that of a conducting material and lower than that of an insulating material. They are small in size and are commonly made using metallic oxides, stabilizers, and binders pressed into disc form. These devices are sensitive to small changes in temperature and its resistance changes accordingly, which can be quantified. Thermistors are of two types: NTC (negative temperature coefficient) type and PTC (positive temperature coefficient) type. NTC thermistors are characterized with a decrease in resistance with rise in temperature and in PTC type resistance increases with temperature rise. The first category is commonly used as temperature sensors.

Thermocouples are the most common sensors for temperature measurement, and it contains two different metallic wires joined at one end to form an electrical junction. Thermal variation at the junction causes generation of a small voltage across the conductors, according to the Seebeck effect, which can be measured, and subsequent temperature values can be interpreted. These devices are simple and inexpensive and are self-powered avoiding the requirement of any external form of excitation. Options for different material combinations make the system versatile enabling measurement at wide ranges. The major limitation while using thermocouples is its lower accuracy.

Resistance temperature detectors (RTDs) work on the same principle as that of thermistors. These devices are more stable and offer measurement over a wide temperature range and at extreme conditions. Commonly platinum, copper, or nickel materials are used as wires as they can offer a positive temperature coefficient. Platinum RTDs are preferred in the industrial sector as they offer superior corrosion resistance, measurement over a wide range of temperature, and long-term stability.

In TCM, temperature monitoring is important as it can enhance chemical dissolution and subsequent tool failure. Several methods are adopted by researchers for on time temperature measurement like inserted thermocouples, tool-work thermocouples, radiation thermography, and thin film thermal sensors, etc. (Hoyne et al. 2015). Thermal imaging using radiation thermography has the advantage of estimating the temperature without making contact with the measurement object. The working is based on the idea of capturing, recording, and analysing the infrared radiation emitted by the object. It is often difficult to get precise data using these devices because of the difficulty in attaching these sensors at the tiny tip of the tool/difficulty in focusing the exact machining zone, where maximum temperature rise may happen. Thus, the signals or data are normally calibrated to estimate temperature at the exact machining zone. Because of this inaccuracy, the temperature data is not always preferred in TCM systems.

11.5 TOOL CONDITION MONITORING IN MECHANICAL MICROMACHINING

Xiaozhi and Beizhi (2007) used an AE sensor for monitoring the condition of the tool (tungsten carbide) while turning of mild steel. The AE signals were collected during machining with sharp and worn-out tools. They found that the dominant frequency of the AE signal is in the range of 120 to 150 kHz. From the literature, it is observed that the peak-to-peak (amplitude) of the AE signal is found to be dependent on the condition of the tool (good or worn out). The dominant frequency of the AE signals varies with different machining conditions (Diniz et al. 1992). In the case of broaching, it is found to be in the range between 110 and 225 kHz (Axinte and Gindy 2003), while in the grinding process, it is found to be in the range of 20 to 120 kHz (Guo and Ammula 2005).

Balan (2008) used different AE parameters such as AERMS, ringdown count, and rise time for monitoring the tool wear while micro turning of OFHC copper with cermet insert. He observed that all the AE parameters increased with progressive rise in tool wear. He concluded that the correlation of AERMS and rise time is useful in monitoring the tool wear. Ranjith et al. (2010) used AE signals to monitor the tool wear in micro turning of copper while machining with cermet inserts. A progressive rise in AERMS signals was observed with an increase in cutting velocity. They observed that a good correlation exists between tool wear and AERMS signals.

Dan and Mathew (1990) reported that the dominant frequency of the vibration signal is found to be in the range of 4 to 8 kHz during turning operation. They observed that this phenomenon is consistent even when there is a change in the input parameters such as speed, feed and depth of cut for a wide range. El-Wardany et al. (1996) measured vibration signals during drilling. The dominant frequency of the vibration signal is found to be in the range between 3 and 5 kHz. Chelladurai et al. (2008) found that the amplitude of vibration signal increases with the increase in depth of cut and feed rate. They have analysed vibration signals collected during the turning operation and found that it is useful in predicting flank wear. From the literature, it is observed that an accelerometer is an efficient sensor which can be used to monitor the condition of a tool and it is found that there is no work reported in the literature for monitoring the tool in micro turning. Hence, an accelerometer is used in the present work for the first time to monitor the condition of the tool.

Grinding is a complex machining process associated with a high level of uncertainties, non-linearities, and time dependent characteristics. Effective tool condition monitoring in grinding is a tedious task as the machining involves several micro-cutting processes exhibited by several randomly oriented abrasive particles held on the surface of the grinding wheel. Thus, many factors can disturb the process like random wear of abrasive particles, elastic deformation, and associated dimensional variation for cutting tool and workpiece, hardness variation due to cyclic thermal stresses, wheel mounting errors, etc. The scenario is much more complicated in the domain of micro-grinding because of the extremely small size of the grinding tool and a minor variation in the process can result in severe damage of both tool and workpiece. Apart from the conventional condition monitoring systems, different approaches were used for process monitoring in micro-grinding.

Lee et al. (2014) developed a model for tool condition monitoring in micro-grinding using wavelet packet decomposition and back-propagation neural network approaches. The average value of the measured tangential force readings was used for defining tool conditions while developing the model. It was observed that the developed model could outperform the existing models through its capability in applying to multiple grinding cases with varying feed rate and depth of cut. Feng et al. (2009) studied the influence of parameters like tool stiffness and tool wear on grinding process signals like force, acoustic emission, vibration, spindle load, etc. They observed that prediction of tool wear from the above signals is a tedious task as the signals exhibited a random variation as the tool wears. This can be attributed to the substantial tool deflection caused by the low stiffness of the micro-grinding wheel. To overcome this difficulty, a new parameter was introduced combining force and vibration signals on the idea of variable cutting stiffness. Using this approach experiments were conducted, and they established the fact that tool wear monitoring is possible in micro-grinding even without the information on machining characteristics.

11.6 TOOL CONDITION MONITORING USING MULTIPLE SENSORS

Researchers have developed different types of sensors for TCM systems; however, a single sensor for monitoring the tool condition is not reliable for a wide range of applications due to their insufficient sensitivity at different operating conditions and machining environments. Therefore, researchers use multiple sensors to collect the maximum amount of information about the state of the process. The integration of information from different sensors will increase the accuracy and reliability in monitoring manufacturing processes for decision making. Many researchers used multiple sensors for monitoring the tool condition in micromachining and in micro turning

They concluded that a multi-sensor approach leads to reliable results. From their studies, it is observed that the integration of different sensors for TCM systems is necessary to enhance productivity, quality, and reliability. Hence in the present work, TCM during micro turning has been carried out using multiple sensors such as AE and accelerometer sensors simultaneously to monitor the micro tool during micro turning.

11.6.1 Advantages of tool condition monitoring

Tool condition monitoring has completely changed the existing traditional maintenance concepts such as breakdown and time based preventive maintenance. Reduction in downtime is one of the primary advantages of using proper tool condition monitoring systems which ultimately result in the reduction in maintenance, material wastage, and total manufacturing cost (Fogliazza et al. 2021; Zhou et al. 2022). TCM also eliminates secondary damage to machines and enables the freedom to control dimensional accuracy and surface quality even in mass production. A properly used TCM can reveal the details of tool wear progress, chatter, chip condition, machining temperature, and premature tool wear apart from the dimensional accuracy and surface finish (Nath 2020). A well-developed TCM makes the machines Industry 4.0 ready.

11.7 RESEARCH DIRECTION

In the present modern semi-autonomous and autonomous production industries, an efficient, sustainable, and reliable tool condition monitoring system has been of significant interest. It is crucial to make the necessary corrections in the process anomaly to detect the tool wear or failure accurately and quickly. Presently, with the growth of various sensors with data/signal acquisition methods, TCM is growing at a rapid phase. More research and development are focused on implementing artificial intelligence techniques using the data from TCM. This can lead to creating an intelligent machine tool. Research is also directed towards the availability and feasibility of relevant hardware and software systems.

11.8 CONCLUSION

Tool wear is a critical problem in most of the manufacturing industries as it has adverse effects on product quality, production rate, and working environment. Since the manufacturing sector is now focused on sustainable production, it is important to monitor tool wear continuously during the machining process. However, a direct measurement can adversely affect the production rate as it increases the idle time. Thus, an indirect method like online tool condition monitoring is always preferred as it makes the measurement process much easier. Further it has the added advantage of easiness in installation also. However, the effectiveness of such systems and the performance of the machining process depends on the selection of adequate sensors having essential features. The present chapter is intended to give an overall picture of different sensors that can be adopted in tool condition monitoring capturing signals like vibration, cutting force, temperature, AE, torque, etc., particularly in the micromachining domain. Along with that a brief description of some of the recent research works which incorporated such sensors for tool condition monitoring in different micromachining processes are also included. These studies point towards the significant improvements in process efficiency with the addition of online monitoring systems in the machining process. However, process monitoring through direct incorporation sensors was not that easy and effective in micro-grinding because of the difficulty in accessing the machining zone. Different approaches were tried in grinding to overcome such difficulties.

REFERENCES

Alhadeff, L., M. Marshall, D. Curtis and T. Slatter (2020) Applying experimental micro-tool wear measurement techniques to industrial environments. *Proceedings of the IMechE Part B: Journal of Engineering Manufacture*, 235(10), 1588–1601.

Alting, L., F. Kimura, H.N. Hansen and P. Bissacco (2003) Micro engineering. *Annals of CIRP*, 52(2), 635–657.

Arif, M., M. Rahman and W.Y. San (2012) An experimental investigation into micro ball end-milling of silicon. *Journal of Manufacturing Processes*, 14(1), 52–61.

Asad, A.B.M.A., T. Masaki, M. Rahman, H.S. Lim and Y.S. Wong (2007) Tool-based micro-machining. *Journal of Materials Processing Technology*, 192–193, 204–211.

Axinte, D.A. and N. Gindy (2003) Tool condition monitoring in broaching. *Wear*, 254(3–4), 370–382.

Balan, A.S.S. (2008) Process monitoring during micromachining of copper. MS by Research. Thesis, Anna University, Chennai.

Ballantine, D.S., R.M. White, S.J. Martin, A.J. Ricco, E.T. Zellers, G.C. Frye and H. Wohltjen (1997) *Acoustic Wave Sensors: Theory, Design and Physic-chemical Applications*, Academic Press, San Diego, USA

Bao, W.Y. and I.N. Tansel (2000a) Modeling micro-end-milling operations. Part I: Analytical cutting force model. *International Journal of Machine Tools and Manufacture*, 40(15), 2155–2173.

Bao, W.Y. and I.N. Tansel (2000b) Modeling micro-end-milling operations. Part II: Tool run-out. *International Journal of Machine Tools and Manufacture*, 40(15), 2175–2092.

Botsaris, P.N. and J.A. Tsanakas (2008) State-of-the-art in methods applied to tool condition monitoring (Tcm) in unmanned machining operations: A review. In *Proceedings of the International Conference of COMADEM*, 11-13 June 2008, Preditest/Czech Society of NDT, Prague, Czech Republic, 73–87.

Byrne, G., D. Dornfeld, I. Inasaki, G. Ketteler, W. Konig and R. Teti (1995) Tool condition monitoring (TCM) – The status of research and industrial application. *Annals of CIRP*, 44(2), 541–567.

Chae, J., S.S. Park and T. Freiheit (2005) Investigation of micro-cutting operations. *International Journal of Machine Tools and Manufacture*, 20, 1–20.

Chelladurai, H., V.K. Jain and N.S. Vyas (2008) Development of a cutting tool condition monitoring system for high speed turning operation by vibration and strain analysis. *International Journal of Advanced Manufacturing Technology*, 37(5–6), 471–485.

Dan, L. and J. Mathew (1990) Tool wear and failure monitoring techniques for turning - A review. *International Journal of Machine Tools and Manufacture*, 30(4), 579–598.

Davis, C.E. (1971) A new dynamometer principle. *International Journal of Machine Tool Design and Research*, 11(1), 31–43.

Diniz, A.E., J.J. Liu and D.A. Dornfeld (1992) Correlating tool life, tool wear and surface roughness by monitoring acoustic emission in finish turning. *Wear*, 152(2), 395–407.

Dornfeld, D., S. Min and Y. Takeuchi (2006) Recent advances in mechanical micro-machining. *Annals of CIRP*, 55(2), 745–768.

Dornfeld, D.A. and D.E. Lee (2008) *Precision Manufacturing*, Springer, New York.

Dow, T.A., E.L. Miller and K. Garrard (2004) Tool force and deflection compensation for small milling tools. *Precision Engineering*, 28(1), 31–45.

El-Wardany, T.I., D. Gao and M.A. Elbestawi (1996) Tool condition monitoring in drilling using vibration signature analysis. *International Journal of Machine Tools and Manufacture*, 36(6), 687–711.

Escolle, B., M. Fontaine, A. Gilbin, S. Thibaud and P. Picart (2015) Experimental investigation in micro ball-end milling of hardened steel. *Journal of Materials Science and Engineering*, A5(9–10), 319–330.

Fard, M.J.B. and E.V. Bordatchev (2013) Experimental study of the effect of tool orientation in five-axis micro-milling of brass using ball-end mills. *International Journal of Advanced Manufacturing Technology*, 67(5–8), 1079–1089.

Feng, J., B.S. Kim, A. Shih and J. Ni (2009) Tool wear monitoring for micro-end grinding of ceramic materials. *Journal of Materials Processing Technology*, 209(11), 5110–5116.

Fogliazza, G., C. Arvedi, C. Spoto, L. Trappa, F. Garghetti, M. Grasso and B.M. Colosimo (2021) Fingerprint analysis for machine tool health condition monitoring. *IFAC-PapersOnLine*, 54–1(1), 1212–1217.

Guo, Y.B. and S.C. Ammula (2005) Real-time acoustic emission monitoring for surface damage in hard machining. *International Journal of Machine Tools and Manufacture*, 45(14), 1622–1627.

Hong, Y.S., H.S. Yoon, J.S. Moon, Y.M. Cho and S.H. Ahn (2016) Tool-wear monitoring during micro-end milling using wavelet packet transform and fisher's linear discriminant. *International Journal of Precision Engineering and Manufacturing*, 17(7), 845–855.

Hoyne, C. and S.G. Nath (2015) Kapoor, On temperature measurement during titanium machining with the atomization-based cutting fluid (ACF) spray system, ASME. *Journal of Manufacturing Science and Engineering*, 137(2), 024502.

Jackson, M.J. (2007) *Micro and Nanomanufacturing*, Springer, New York.

Jain, V.K. (2010) *Introduction to Micromachining*, Narosa Publishing House, New Delhi.

Jemielniak, K. (1999) Commercial tool condition monitoring systems. *The International Journal of Advanced Manufacturing Technology*, 15(10), 711–721.

Jemielniak, K. and P.J. Arrazola (2008) Application of AE and cutting force signals in tool condition monitoring in micro-milling. *CIRP Journal of Manufacturing Science and Technology*, 1(2), 97–102.

Jin, X. and Y. Altintas (2012) Prediction of micro-milling forces with finite element method. *Journal of Materials Processing Technology*, 212(3), 542–552.

Kang, H.J. and S.H. Ahn (2007) Fabrication and characterization of microparts by mechanical micromachining: Precision and cost estimation. *Proceedings of the IMechE Part B: Journal of Engineering Manufacture*, 221(2), 231–240.

Kang, H.J. and S.H. Ahn (2007) Fabrication and characterization of microparts by mechanical micromachining: Precision and cost estimation. *Proceedings of the IMechE Part B: Journal of Engineering Manufacture*, 221(2), 231–240.

Kim, J.D. and D.S. Kim (1995) Theoretical analysis of microcutting characteristics in ultra-precision machining. *Journal of Materials Processing Technology*, 49(3–4), 387–398.

Kistler (2009) *Cutting Force Measurement- Precise Measuring Systems for Metal-Cutting*. Kistler Group. https://www.scribd.com/document/76500100/Dinamometre-Fixe-Si-Rotative. (Accessed 31 October 2021).

Kistler (2014) *Product Catalog- Sensors and Solutions for Cutting Force Measurement*. Kistler Group, https://www.kistler.com/fileadmin/files/divisions/sensor-technology/cutting-force/960-002e-05.14.pdf (Accessed 31 October 2021).

Koc, M. and T. Ozel (2011) *Micro-Manufacturing*, John Wiley and Sons, New Jersey.

Lee, P. H., D.H. Kim, D.S. Baek, J.S. Nam and S.W. Lee (2014) A study on tool condition monitoring and diagnosis of micro-grinding process based on feature extraction from force data. *Proceedings of the Institution of Mechanical Engineers, Part B*, 1–7.

Li, C., X. Lai, H. Li and J. Ni (2007) Modeling of three-dimensional cutting forces in micro-end-milling. *Journal of Micromechanics and Microengineering*, 17(4), 671–678.

Li, C., X. Lai, H. Li and J. Ni (2007) Modeling of three-dimensional cutting forces in micro-end-milling. *Journal of Micromechanics and Microengineering*, 17(4), 671–678.

Li, S. and K. Zhu (2021) In-situ tool wear area evaluation in micro milling with considering the influence of cutting force. *Mechanical Systems and Signal Processing*, 161, 107971.

Liow, J.L. (2009) Mechanical micromachining: A sustainable micro device manufacturing approach. *Journal of Cleaner Production*, 17(7), 662–667.

Malekian, M., S.S. Park and M.B.G. Jun (2009) Tool wear monitoring of micromilling operations. *Journal of Materials Processing Technology*, 209(10), 4903–4914.

Masuzawa, T. and H.K. Tonshoff (1997) Three dimensional micromachining by machine tools. *Annals of CIRP*, 46(2), 621–628.

McGeough, J. (2002) *Micromachining of Engineering Materials*, Marcel Dekker, New York.

Mohanraj, T., S. Shankar, R. Rajasekar, N.R. Sakthivel and A. Pramanik (2020) Tool condition monitoring techniques in milling process — A review. *Journal of Materials Research and Technology*, 9(1), 1032–1042.

Nath, C. (2020) Integrated tool condition monitoring systems and their applications: A comprehensive review. *Procedia Manufacturing*, 48, 852–863.

Ogedengbe, T.I., R. Heinemann and S. Hinduja (2011) Feasibility of tool condition monitoring on micro-milling using current signals. *Journal of Technology*, 14(3), 161–172.

Oliaei, S.N.B. and Y. Karpat (2016) Influence of tool wear on machining forces and tool deflections during micro milling. *International Journal of Advanced Manufacturing Technology*, 84(9–12), 1963–1980.

Ozel, T. and X. Liu (2009) Investigations on mechanics-based process planning of micro-end milling in machining mold cavities. *Materials and Manufacturing Processes*, 24(12), 1274–1281.

Perez, H., A. Vizan, J.C. Hernandez and M. Guzman (2007) Estimation of cutting forces in micromilling through the determination of specific cutting pressures. *Journal of Materials Processing Technology*, 190(1–3), 18–22.

Qin, Y. (2010a) *Micro-manufacturing Engineering and Technology*, Elsevier, USA

Qin, Y., A. Brockett, Y. Ma, A. Razali, J. Zhao, C. Harrison, W. Pan, X. Dai and D. Loziak (2010b) Micro-manufacturing: Research, technology outcomes and development issues. *International Journal of Advanced Manufacturing Technology*, 47(9–12), 821–837.

Rahman, M.A., M. Rahman, A.S. Kumar, H.S. Lim and A.B.M.A. Asad (2006) Development of micropin fabrication process using tool based micromachining. *International Journal of Advanced Manufacturing Technology*, 27(9–10), 939–944.

Rajurkar, K.P., G. Levy, A. Malshe, M.M. Sundaram, J. McGeough, X. Hu, R. Resnick and A. Desilva (2006) Micro and nano machining by electro-physical and chemical processes. *Annals of CIRP*, 55(2), 643–666.

Ranjith, P.K., A.S.S. Balan and S. Gowri (2010) Monitoring and prediction of tool wear in microturning of copper. *International Journal of Precision Technology*, 1(3/4), 343–355.

Ravisubramanian, S. and M.S. Shunmugam (2015) On reliable measurement of micro drilling forces and identification of different phases. *Measurement*, 73, 335–340.

Robinson, G.M. and M.J. Jackson (2005) A review of micro and nanomachining from a materials perspective. *Journal of Materials Processing Technology*, 167(2–3), 316–337.

Rodriguez, P. and J.E. Labarga (2013) A new model for the prediction of cutting forces in micro-end-milling operations. *Journal of Materials Processing Technology*, 213(2), 261–268.

Sahoo, P., K. Patra, V.K. Singh, R.K. Mittal and R.K. Singh (2020) Modeling dynamic stability and cutting forces in micro milling of Ti6Al4V using intermittent oblique cutting finite element method simulation-based force coefficients. *Journal of Manufacturing Science and Engineering*, 142(9), 091005.

Soin, N., S.C. Anand and T.H. Shah (2016) Energy harvesting and storage textiles. In Horrocks A.R. and S.C. Anand (Eds.), *Handbook of Technical Textiles* (2nd ed.), Woodhead Publishing, United Kingdom, 357–396.

Sun, P.Y., Y.K. Chang, T.C. Wang and P.T. Liu (1982) A simple and practical piezo-electric shank type three-component dynamometer. *International Journal of Machine Tool Design and Research*, 22(2), 111–124.

Tacuna Systems (2009) *Comparing Strain Gauges to Piezoelectric Sensors*. Tacuna Systems. https://tacunasystems.com/knowledge-base/comparing-strain-gauges-to-piezoelectric-sensors/ (Accessed 15 October 2021).

Taniguchi, N. (1983) Current status in and future trends of ultraprecision machining and ultrafine materials processing. *Annals of the CIRP*, 32(2), 573–582.

Tansel, I.N., T.T. Arkan, W.Y. Bao, N. Mahendrakar, B. Shisler, D. Smith and M. McCool (2000a) Tool wear estimation in micro-machining Part II: Neural-network-based periodic inspector for nonmetals. *International Journal of Machine Tools and Manufacture*, 40(4), 609–620.

Tansel, T.T., W.Y. Bao Arkan, N. Mahendrakar, B. Shisler, D. Smith and M. McCool (2000b) Tool wear estimation in micro-machining Part II: Neural-network-based periodic inspector for nonmetals. *International Journal of Machine Tools and Manufacture*, 40(4), 609–620.

Teti, R. (1992) Cutting conditions and work material state identification through acoustic emission methods. *Annals of CIRP*, 41(1), 89–92.

Tonshoff, H.K. and I. Inasaki (2001) *Sensors in Manufacturing*, Wiley-Vch Verlag GmbH, Weinheim.

Uriarte, L., S. Azcarate, A. Herrero, L.N. Lopez de Lacalle and A. Lamikiz (2008) Mechanistic modeling of the micro end milling operation. *Proceedings of the IMechE Part B: Journal of Engineering Manufacture*, 222(1), 23–33.

Wang, L. and R.X. Gao (2006) *Condition Monitoring and Control for Intelligent Manufacturing*, Springer-Verlag, London.

Wojciechowski, S. (2015) The estimation of cutting forces and specific force coefficients during finishing ball end milling of inclined surfaces. *International Journal of Machine Tools and Manufacture*, 89, 110–123.

Xiaoli, L. "A brief review: acoustic emission method for tool wear monitoring during turning", *International Journal of Machine Tools and Manufacture*, Vol. 42, pp. 157–165, 2002.

Xiaozhi, C. and L. Beizhi (2007) Acoustic emission method for tool condition monitoring based on wavelet analysis. *International Journal for Advanced Manufacturing Technology*, 33, 968–976.

Xu, C., J. Zhu and S.G. Kapoor (2015) Force modeling of five-axis micro ball-end milling. *Journal of Micro- and Nano-Manufacturing*, 3, 1–13, 031007.

Yeo, S.H., L.P. Khoo and S.S. Neo (2000) Tool condition monitoring using reflectance of chip surface and neural network. *Journal of Intelligent Manufacturing*, 11(6), 507–514.

Zhang, X., T. Yu, P. Xu and J. Zhao (2022) In-process stochastic tool wear identification and its application to the improved cutting force modeling of micro milling. *Mechanical Systems and Signal Processing*, 164, 108233.

Zhou, Y., G. Zhi, W. Chen, Q. Qian, D. He, B. Sun and W. Sun (2022) A new tool wear condition monitoring method based on deep learning under small samples. *Measurement*, 189, 110622.

Index

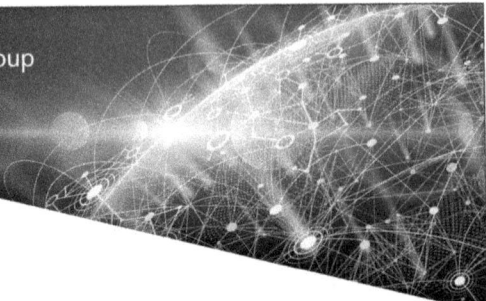